Springer Theses

Recognizing Outstanding Ph.D. Research

Aims and Scope

The series "Springer Theses" brings together a selection of the very best Ph.D. theses from around the world and across the physical sciences. Nominated and endorsed by two recognized specialists, each published volume has been selected for its scientific excellence and the high impact of its contents for the pertinent field of research. For greater accessibility to non-specialists, the published versions include an extended introduction, as well as a foreword by the student's supervisor explaining the special relevance of the work for the field. As a whole, the series will provide a valuable resource both for newcomers to the research fields described, and for other scientists seeking detailed background information on special questions. Finally, it provides an accredited documentation of the valuable contributions made by today's younger generation of scientists.

Theses are accepted into the series by invited nomination only and must fulfill all of the following criteria

- They must be written in good English.
- The topic should fall within the confines of Chemistry, Physics, Earth Sciences, Engineering and related interdisciplinary fields such as Materials, Nanoscience, Chemical Engineering, Complex Systems and Biophysics.
- The work reported in the thesis must represent a significant scientific advance.
- If the thesis includes previously published material, permission to reproduce this must be gained from the respective copyright holder.
- They must have been examined and passed during the 12 months prior to nomination.
- Each thesis should include a foreword by the supervisor outlining the significance of its content.
- The theses should have a clearly defined structure including an introduction accessible to scientists not expert in that particular field.

More information about this series at http://www.springer.com/series/8790

Xin Wu

Influence of Particle Beam Irradiation on the Structure and Properties of Graphene

Doctoral Thesis accepted by
Tsinghua University, Beijing, China

Author
Dr. Xin Wu
Department of Mechanical Engineering
Tsinghua University
Beijing
China

Supervisor
Prof. Haiyan Zhao
Department of Mechanical Engineering
Tsinghua University
Beijing
China

ISSN 2190-5053 ISSN 2190-5061 (electronic)
Springer Theses
ISBN 978-981-10-6456-2 ISBN 978-981-10-6457-9 (eBook)
https://doi.org/10.1007/978-981-10-6457-9

Library of Congress Control Number: 2017954480

© Springer Nature Singapore Pte Ltd. 2018
This work is subject to copyright. All rights are reserved by the Publisher, whether the whole or part of the material is concerned, specifically the rights of translation, reprinting, reuse of illustrations, recitation, broadcasting, reproduction on microfilms or in any other physical way, and transmission or information storage and retrieval, electronic adaptation, computer software, or by similar or dissimilar methodology now known or hereafter developed.
The use of general descriptive names, registered names, trademarks, service marks, etc. in this publication does not imply, even in the absence of a specific statement, that such names are exempt from the relevant protective laws and regulations and therefore free for general use.
The publisher, the authors and the editors are safe to assume that the advice and information in this book are believed to be true and accurate at the date of publication. Neither the publisher nor the authors or the editors give a warranty, express or implied, with respect to the material contained herein or for any errors or omissions that may have been made. The publisher remains neutral with regard to jurisdictional claims in published maps and institutional affiliations.

Printed on acid-free paper

This Springer imprint is published by Springer Nature
The registered company is Springer Nature Singapore Pte Ltd.
The registered company address is: 152 Beach Road, #21-01/04 Gateway East, Singapore 189721, Singapore

Supervisor's Foreword

It is my pleasure to introduce Dr. Xin Wu's original work to be published in the series of Springer Theses, which is mainly about the using of particle beam irradiation to process graphene and the investigation of the properties of the obtained graphene nanostructures.

Graphene is a one-atom-thick plane sheet of sp^2-bonded carbon atoms tightly packed into a honeycomb crystal lattice. As a true two-dimensional material newly discovered, graphene behaves extraordinary properties, i.e., outstanding mechanical properties and electrical properties, unique optical properties, high thermal properties, etc., which promise graphene numerous potential applications in different areas. In order to achieve these potential applications, graphene needs to be processed into nanostructures. As for the nanomanufacture of graphene, on the one hand, we need to join different sheets to effectively govern the size and shape of single graphene sheet. On the other hand, we need to dope graphene to open its zero-energy band and master the electrical properties. Meanwhile, the fabrication of graphene nanopore is desiderated in nanopore sequence, desalination, etc. Nevertheless, most methods have shortages in the nanomanufacture of graphene. Nanomanufacturing by particle beams shows obvious advantages over the conventional physicochemical methods. Therefore, this thesis proposed the using of particle beam irradiation method to dope graphene, joining graphene, and fabricate graphene nanopore. The processing mechanisms and the properties of the fabricated graphene nanostructures are analyzed by experiments and atomic simulations.

The first part of this work is to figure out the interaction mechanisms between particle beam and graphene under laser beam, ion beam, and electron beam irradiation. The damage threshold of graphene under particle beam irradiation was discovered, as well as the initiation and evolution of defects in graphene structure. Meanwhile, the processing precision of the nanostructure was analyzed, and its influence factors were discussed.

Based on the above interaction mechanisms, this thesis experimentally proved the feasibility of graphene doping by low-energy nitrogen irradiation and explained the doping mechanism by atomistic simulation: The low-energy ion beam irradiation will first result in many adatoms and vacancies in graphene, and then these adatoms will migrate and combine with the vacancies during the equilibrium process, which would result in substitution doping. If the adatoms cannot completely fill in the vacancies, there will be many polygon defects alongside the substituted atom. The thesis also explored the influence of aggregation, the defects, and the type of doping elements on the mechanical properties of graphene. In addition, the electronic transport properties of doped graphene were studied in consideration of the influence of doping type and doping site.

Afterward, this thesis experimentally proved that ion beam irradiation and laser can join two overlapped graphene sheets. For the ion beam irradiation case, different layers of graphene can form new bonds, which are ascribed to two mechanisms: saturation of carbon atoms in graphene and the embedded ions induced bridging. For the laser irradiation and annealing case, the joining is attributed to the enhanced intermolecular forces. It was found that there are always defects in the butt joint, which easily raise concentration of stress during stretch, and these defects are determined by the relative deflection angle of two sheets. The thesis discovered the relationship between the mechanical properties and the irradiated ion dose, energy, and type and proposed that the electronic transport properties of the butt joint are heavily blocked by the difference of individual orbital energy and localized state induced by defects.

Additionally, the thesis used FIB and electron beam to fabricate nanopore in graphene. The mechanisms located in nanopore fabrication were illustrated, i.e., the nanopore in suspended graphene is fabricated due to the "direct sputtering" induced by irradiated ions and the "neighbor dragging" induced by sputtered atoms, while the nanopore in supported graphene is developed because of "direct sputtering," "neighbor dragging," and the "indirect sputtering" induced by substrate atoms. After that, the influence of nanopore size, defects, and chirality of graphene was taken into consideration to research the mechanical and electronic transport properties of graphene nanopore.

The conclusions of his work can help to promote the realization of the applications of graphene nanostructures in industry. Part of this work has already been published in high-profile journals such as *Carbon* and *Applied Physics Letters*. I believe that the full publication of this thesis in Springer will further promote the research in the field of graphene material.

Beijing, China   Prof. Haiyan Zhao
July 2017

Parts of this thesis have been published in the following journals:

1. Xin Wu; Haiyan Zhao; Jiayun Pei; Dong Yan. Joining of graphene flakes by low energy N ion beam irradiation. *Appl. Phys. Lett.* **2017**, *110*, 133102.

2. Xin Wu; Haiyan Zhao; Dong Yan; Jiayun Pei. Doping of graphene using ion beam irradiation and the atomic mechanism. *Comp. Mater. Sci.* **2017**, *129*, 184–193.

3. Xin Wu; Haiyan Zhao; Jiayun Pei. Fabrication of nanopore in graphene by electron and ion beam irradiation: influence of graphene thickness and substrate. *Comp. Mater. Sci.* **2015**, *102*, 258–266.

4. Xin Wu; Haiyan Zhao; Dong Yan; Jiayun Pei. Investigation of gallium ions impacting monolayer graphene. *AIP Adv.* **2015**, *5*, 067171.

5. Xin Wu; Haiyan Zhao; Minlin Zhong; Hedekazu Murakawa; Masahiro Tsukamoto. Molecular dynamics simulation of graphene sheets joining under ion beam irradiation. *Carbon* **2014**, *66*, 31–38.

6. Xin Wu; Haiyan Zhao; Hedekazu Murakawa. The joining of graphene sheets under Ar ion beam irradiation. *J. Nanosci. Nanotechno.* **2014**, *14*, 5697–5702.

7. Xin Wu; Haiyan Zhao; Minlin Zhong, et al. The formation of molecular junctions between graphene sheets. *Mater. Trans.* **2013**, *54*, 940–946.

Acknowledgements

This thesis cannot be finished without many people's help. I would like to express my great gratitude to all those who provided support.

Firstly, I would like to express my sincere gratitude and deep respect to Prof. Haiyan Zhao. The work of this thesis is completed under the guidance of Prof. Zhao. As a fantastic, knowledgeable, rigorous supervisor, Prof. Zhao contributed his great help and support to my research work and life. His magnificent insight and educational effort also have an important impact on my future life.

Many thanks to Prof. Hongwei Zhu, for his constructive guidance and support during the topic selection of this thesis and the conduct of the experimental work.

My sincere thanks to Prof. Hedekazu Murakawa, Prof. Ninshu Ma, and Prof. Masahiro Tsukamoto. During the six-month collaborative study in Osaka University, you offered me a lot of guidance and help, thank you very much.

The completion of this paper is inseparable from the care and support of the group members including Dong Yan, Pu Xie, Lugui He, Jiankun Xiong, Junwei Wu, Yue Liu, Jiayun Pei, Gaoqiang Chen, Han Li, Qu Liu, Xiong Cao, Yucan Zhu, and other students. Thank you so much to all of you.

This paper is funded by the Beijing Natural Science Foundation (3142010) program, the Specialized Research Fund for the Doctoral Program of Higher Education of China (20130002110088), Tsinghua University Initiative Scientific Research Program (20161080170), and Fundings of State Key Lab of Tribology in Tsinghua University (SKLT2014A03). Thank you for the financial support.

Contents

1	**Introduction**		1
	1.1 Basic Introduction of Graphene		1
		1.1.1 Overview of the Development History of Graphene	1
		1.1.2 Structure and Properties of Graphene	3
		1.1.3 Main Preparation Methods of Graphene	6
		1.1.4 Applications of Graphene	8
	1.2 Processing of Graphene		10
		1.2.1 Joining of Graphene	10
		1.2.2 Fabrication of Graphene Nanopore	11
		1.2.3 Doping of Graphene	11
	1.3 Particle Beam Processing and Its Application in Graphene Structure		13
		1.3.1 Introduction of Particle Beam Processing	13
		1.3.2 Research Status of Particle Beam Processing Graphene	14
	1.4 Problem Introduction		16
	1.5 Research Content		18
	References		19
2	**Experiment Approaches and Simulation Methods**		23
	2.1 Synthesis and Characterization of Graphene Specimen		23
		2.1.1 Preparation of Monolayer and Multilayer Graphene Specimens	23
		2.1.2 The Main Characterization Methods of Graphene Sample	26
		2.1.3 The Main Experimental Equipment for Graphene Processing	28
	2.2 Introduction of MD Simulation		29
		2.2.1 Concepts	29
		2.2.2 Basic Principles of Classic MD	29
		2.2.3 Atomic Interaction Force	30
		2.2.4 Integral Algorithm	34

	2.2.5	Simulation Ensemble.	35
	2.2.6	Averaging of Statistical Results.	36
	2.2.7	Introduction of Simulation Software	38
2.3	Electronic Transport Theory.		38
	2.3.1	Introduction.	38
	2.3.2	DFT	38
	2.3.3	Green Function Theory	43
	2.3.4	Solution Process of the Electronic Transport Properties of Graphene	47
	2.3.5	Introduction to Simulation Software	48
2.4	Chapter Summary.		48
References.			50

3 General Mechanisms During the Interaction Between Particle Beam and Graphene ... 51

3.1	Introduction		51
3.2	Interaction Between Laser Beam and Graphene		52
	3.2.1	Damage Threshold of Graphene Irradiated by Single Pulse Laser	52
	3.2.2	The Change of the Morphology of Graphene Under Ultrafast Laser.	54
	3.2.3	Experimental Processing of Graphene Structure Under Ultrafast Laser Irradiation	56
3.3	Interaction Between Ion Beam and Graphene.		57
	3.3.1	The Phenomenon of Graphene Irradiated by Different Energy Ion Beam	57
	3.3.2	Effect of Substrate on Ion Beam Irradiation of Graphene	60
3.4	Interaction Between Electron Beam and Graphene		64
	3.4.1	Experimental Study on the Change of Graphene Structure by Electron Beam Irradiation	65
	3.4.2	Mechanism of the Destruction of Graphene Structure Under Electron Beam Irradiation.	67
3.5	Chapter Summary.		71
References.			72

4 Doping of Graphene Using Low Energy Ion Beam Irradiation and Its Properties. ... 73

4.1	Introduction		73
4.2	Experimental Studies of Graphene Doping by Ion Beam Irradiation.		73
	4.2.1	Experiment Procedure.	73
	4.2.2	Experiment Results of Low Energy Ion Implantation Doping	74
	4.2.3	Summary of the Experiment	78

	4.3	Theoretical Analysis of the Doping Mechanism	78
		4.3.1 Research Model	79
		4.3.2 Variation of Graphene Structure Under Nitrogen Ion Implantation	79
		4.3.3 The Energy of the System Corresponding to the Different Doping Configurations	80
		4.3.4 Influence of the Energy and Dose of Implanted Ion Beam	83
	4.4	Mechanical Properties of Doped Graphene by Ion Beam Irradiation	84
		4.4.1 MD Simulation Model	85
		4.4.2 Effect of Implantation Doping on Tensile Stress Distribution of Graphene	86
		4.4.3 The Influence of Doping Concentration and Doping Ion Distribution	87
		4.4.4 Influence of Defect Concentration and Doping Element Type	89
	4.5	Electronic Transport Properties of Doped Graphene	90
		4.5.1 Research Model	91
		4.5.2 Electronic Transport Properties Under Different Doping Forms	92
		4.5.3 Effect of Doping Position on Electrical Performance	95
	4.6	Summary	97
	References		98
5	**Joining of Graphene by Particle Beam Irradiation and Its Properties**		99
	5.1	Introduction	99
	5.2	Experimental Studies of Graphene Joining by Particle Beam Irradiation	99
		5.2.1 Preparation and Characterization of Graphene Joining Samples	100
		5.2.2 Graphene Joining by Ion Beam Irradiation	104
		5.2.3 Graphene Joining by Laser Irradiation and Thermal and Annealing	106
		5.2.4 Experiment Summary	110
	5.3	Theoretical Analysis of the Joining Mechanism	110
		5.3.1 Graphene Joining Under Laser Beam Irradiation	111
		5.3.2 Joining of Graphene by Ion Beam Irradiation	116
	5.4	Mechanical Properties of the Graphene Joint	120
		5.4.1 Mechanical Properties of Butt Joint of Graphene	120
		5.4.2 Mechanical Properties of Overlapped Graphene Joint	125
		5.4.3 Mechanical Properties of Butt Joint Constituted by Multi Pieces of Graphene	129

	5.5	Electric Transport Properties of the Graphene Joint	131
		5.5.1 Electronic Transport Properties of Butt Joint	131
		5.5.2 Electronic Transport Properties of Overlap Joint	136
	5.6	Chapter Summary	140
	References		141
6	**Fabrication of Graphene Nanopore by Particle Beam Irradiation and Its Properties**		143
	6.1	Introduction	143
	6.2	Experimental Studies of Fabrication of Graphene Nanopore by Particle Beam Irradiation	144
		6.2.1 Experiment Procedure	144
		6.2.2 Morphology Analysis of Graphene Nanopore	145
		6.2.3 Effect of Ion Beam Dose on the Properties of Graphene Nanopore	148
		6.2.4 Summary of the Experiment	149
	6.3	Theoretical Analysis of the Fabrication Mechanism of Graphene Nanopore	149
		6.3.1 Research Model	149
		6.3.2 Processing Mechanism of Nanopore	152
		6.3.3 Influencing Factors of Nanopore Processing	155
	6.4	Mechanical Properties of Graphene Nanopore	160
		6.4.1 Research Model	160
		6.4.2 Tensile Failure Process of Graphene Nanopore	161
		6.4.3 Effect of Graphene Chirality on Dynamic Failure Process	162
		6.4.4 Effect of Nanopore Size on Mechanical Properties	164
		6.4.5 Effect of Vacancy Defect on Mechanical Properties of Nanopore	165
	6.5	Electronic Transport Properties of Graphene Nanopore	169
		6.5.1 Research Model	169
		6.5.2 Effect of Nanopore Size on Electrical Properties	170
		6.5.3 Effect of Vacancy Defect on Electrical Performance	172
		6.5.4 Effect of Nanopore Shape on Electrical Properties	173
	6.6	Summary	175
	References		175
7	**Conclusion**		179
	7.1	The Main Conclusions	179
	7.2	Future Plan	182

Abbreviations

AFM	Atomic force microscope
AIREBO	Adaptive intermolecular reactive empirical bond order
CNT	Carbon nanotube
CVD	Chemical vapor deposition
DFT	Density functional theory
FIB	Focused ion beam
GGA	Generalized gradient approximation
HOMO	Highest occupied molecular orbital
HOPG	Highly oriented pyrolytic graphite
LAMMPS	Large-scale Atomic/Molecular Massively Parallel Simulator
LDA	Local density approximation
L-J	Lennard-Jones
LUMO	Lowest unoccupied molecular orbital
MD	Molecular dynamics
NEGF	Non-equilibrium Green's function
NPG	Nanoporous graphene
NPT	Isothermal and isobaric ensemble
NVE	Micro-regular ensemble
NVT	Regular ensemble
PKA	Primary knock-on atom
PMMA	Polymethyl methacrylate
SEM	Scanning electron microscope
TEM	Transmission electron microscope
XPS	X-ray photoelectron spectroscope
Z-A-Z	Zigzag–Armchair–Zigzag
ZBL	Ziegler–Biersack–Littmark
ZGNR	Zigzag graphene nanoribbon

Chapter 1
Introduction

1.1 Basic Introduction of Graphene

1.1.1 Overview of the Development History of Graphene

Carbon is one of the most important constituent elements of nature. It forms the colorful life organism on the earth. As the IVA element in the periodic table, carbon has a $1s^2 2s^2 p^2$ atomic orbital distribution, which leads to the existence of carbon in a variety of structural forms by different hybridization modes (sp, sp^2, sp^3). In terms of elementary substance, carbon is the only one that was found to have stable structures from zero to three dimensions [1]. In fact, people have long time been in contact with a variety of forms of carbon elementary substances in nature: such as diamond, graphite, amorphous carbon. With the development and innovation of science and technology, nano-materials continue to receive attention from the scientific community, and the new carbon nano elementary substances have continually been found. For example, in 1985 a British scientist Harold Kroto, American scientist Robert Curl and Richard Smalley first prepared the fullerene (C60) [2], which is constituted by 60 carbon atoms to get a "football" molecule structure with 12 isolated regular pentagons and 20 isolated regular hexagons. As a new zero-dimensional carbon allotrope, the discovery of fullerene led Harold Kroto et al. to the 1996 Nobel Prize in Chemistry. Then, in 1991 a Japanese researcher Iijima [3] discovered another new type of carbon allotrope-CNT, which is made by crimping graphite film into a one-dimensional carbon allotrope structure. The discovery of CNT also set off a research boom in the field of nano-materials.

With the progress of science and technology, zero-dimensional (fullerene), one-dimensional (CNTs) and three-dimensional (graphite) carbon materials are gradually found and prepared. Does the stable two-dimensional carbon material (i.e. graphene) exist? Before 2004, there was some theoretical work on two-dimensional graphene, but it was generally believed that due to thermodynamic instability, two-dimensional crystal material at room temperature and pressure is easy to break down [4], so the two-dimensional graphene material cannot be stably existed. It was not until 2004 that Geim and Novoselov from the University of Manchester, England, obtained a stable presence of graphene by mechanical exfoliation of HOPG [5]. The graphene material gained great concern again in the scientific community. Scholars also received the 2010 Nobel Prize in Physics for the "Groundbreaking experiment in the study of graphene in two-dimensional materials".

Graphene breaks the Nobel Prize awards record for the period from the discovery to be awarded. Then what is graphene? Boehm gave the earliest definition of graphene in 1986: The term graphene layer should be used for such a single carbon layer [6–8]. Subsequently, the International Theory (Chemistry) and Applied Chemistry Association clearly defined the graphene: the term graphene should be used only when the reactions, structural relations or other properties of individual layers are discussed [9]. In 2014, China's graphene standards committee made a definition of monolayer graphene: monolayer graphene is a two-dimensional carbon material formed by a layer of cyclically closely arranged carbon atoms with benzene ring structure (i.e., hexagonal honeycomb structure) [10]. As a quasi-two-dimensional carbon material composed of a single layer of carbon atoms, graphene can be stably existed by its own surface micro-folds. It forms the basic structural units of other carbon nanomaterials. As shown in Fig. 1.1, it can form zero-dimensional fullerenes by wrapping, one-dimensional CNT by crimping and three-dimensional graphite by stacking [11].

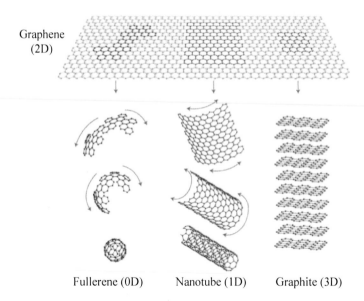

Fig. 1.1 Graphene and carbon isomers formed from graphene. Reprinted by permission from Nature Publishing Group: Ref. [11], Copyright 2007

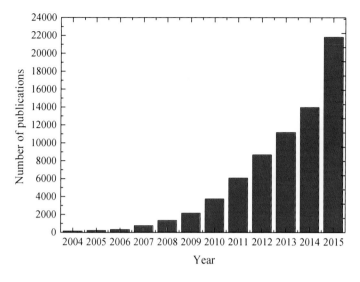

Fig. 1.2 Recent publications on graphene-related research (results from Scopus database)

Due to its special properties such as mechanical, electrical, optical and thermal properties and great potential applications, graphene has been widely concerned by the scientific community since its discovery, which has caused the third research boom of carbon nanomaterials (the first two are corresponding to fullerene and CNT). Figure 1.2 shows the publications of graphene research after 2004. At present, the research hotspots of graphene mainly focus on how to obtain large-scale high-quality graphene, the modification of graphene structure and the application of graphene. Meanwhile, the processing of graphene nanostructure has received more and more attention.

1.1.2 Structure and Properties of Graphene

1.1.2.1 Structure of Graphene

Graphene is essentially a single layer of graphite sheet. Carbon atoms are tightly arranged by sp^2 hybridization into a hexagonal honeycomb structure [12], which is the basic unit of fullerene, CNT and graphite. Graphene is the thinnest of the known materials. It has a thickness of only 0.335 nm, with a carbon-carbon bond length of about 0.142 nm and an angle of 120°. In the lattice plane, each carbon atom forms an σ bond in the form of sp^2 hybridization with three adjacent carbon atoms. The strong covalent bond ensures a stable connection between adjacent atoms. At the same time, the remaining electrons located on the p orbit form a large π bond perpendicular to the crystal plane, and π electrons are free to move in the plane, so that the graphene has very high carrier mobility [13].

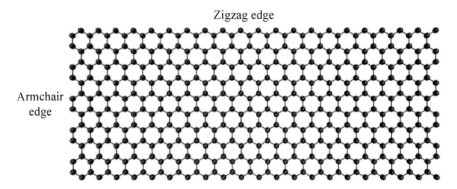

Fig. 1.3 Graphene nanoribbon with chirality

The actual graphene is not a perfectly ideal two-dimensional planar structure. In order to ensure the thermodynamic stability, the monolayer graphene in the experiment has some folds and undulations [14]. In addition, the transfer of graphene to the substrate will produce a lattice to match the deformation of the substrate, resulting in a graphene structure with curved surface, and produce certain internal strain, affecting its performance. Therefore, there is some difference between the actual measured graphene performance and the ideal situation.

Due to the hexagonal lattice structure, graphene has different atomic arrangement along different lattice directions, and its structure is divided into zigzag and armchair according to the carbon chain of graphene edge, as shown in Fig. 1.3. Graphene nanoribbons with different chirality have quite different properties, such as zigzag-type graphene nanoribbon is generally metallic, while armchair-type graphene nanoribbon can be metallic or semiconducting based on its width.

1.1.2.2 Properties of Graphene

The unique two-dimensional structure of the graphene and the atomic orbital bonding method make it unparalleled in other materials. In recent years, the unique properties of graphene in mechanical, electrical, optical, thermal and other aspects were gradually excavated.

1. Mechanical properties

Each carbon atom in graphene forms a stable hexagonal crystal structure with three adjacent carbon atoms through the σ bond. The strong σ bond brings graphene excellent mechanical properties and structural rigidity. In 2008, Lee et al. [15] studied the mechanical properties of suspended graphene by nanoindentation experiments. The tensile strength and elastic modulus of graphene were found to be 130 GPa and 1.0 TPa, respectively. Frank [16] confirmed good elastic properties of graphene and found that the effective spring constant for the graphene with five layers or less ranged 1–5 N/m. Graphene also has a high hardness. Besides, it can be used as reinforcement phase in the composite material to change the performance of other materials [17].

2. Electrical properties

A large π band is formed by the electrons from p atomic orbital when the σ bond is hybrid to be formed in graphene. The π-band electrons are free to move in the crystal plane so that the graphene has extremely high carrier mobility and excellent conductivity. It is confirmed that the carrier mobility in graphene is as high as 2×10^5 cm^2/(Vs) [18], about 140 times of the electron mobility in silicon, and the corresponding conductivity is up to 10^6 S/m [19]. The charge carrier in graphene is Dirac fermion whose ballistic transport has the quantum Hall effect at room temperature [20, 21]. The bandgap of the ideal graphene is zero, and its conduction band and valence band intersect at Dirac's point, as shown in Fig. 1.4. The band structure of graphene can be controlled by chemical doping or processing into different nanostructures to obtain various graphene-based derivatives. For example, if the carbon atoms on the surface of the graphene are bonded to hydrogen atoms, the graphene can be converted to an insulating graphite alkane [22].

3. Optical properties

Graphene has special optical properties. The transmittance of graphene to different visible light is different by 2.3% for every one layer difference [24], so that the layer information can be analyzed by the translucency of graphene. The translucency of graphene can be adjusted by applying a gate voltage or charge injection method. It can be used in optoelectronic devices such as IR detectors, modulators and emitters, according to its optically adjustable property [25]. In addition, graphene presents a nonlinear saturation absorption on light [26, 27], which makes it potentially useful in ultrafast photonics.

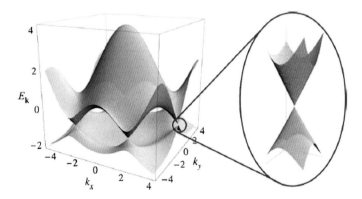

Fig. 1.4 Band structure of graphene. Reprinted with permission from Ref. [23]. Copyright 2009 by American Physical Society

4. Thermal properties

Graphene has good thermal conductivity. The thermal conductivity of graphene at room temperature is about 5×10^3 W/Mk [28], which is 10 times higher than the room temperature thermal conductivity (401 W/mK) of copper. In 2008, Balandin et al. [29] demonstrated that the thermal conductivity of monolayer graphene at room temperature was about $(4.84 \pm 0.44) \times 10^3$ to $(5.30 \pm 0.48) \times 10^3$ W/mK, which is 1 times higher than that of diamond (with highest thermal conductivity in the nature, about 2200 W/mK). The excellent thermal conductivity of graphene makes it potentially valuable in terms of heat dissipation applications.

In addition, graphene also has good biocompatibility [30], ultra-high specific surface area [31] and abundant electrochemical modifiable properties [32]. These special properties make it possible in the applications of every walk of life.

1.1.3 Main Preparation Methods of Graphene

In order to prepare the graphene materials with reliable quality and excellent performance, domestic and international scholars have carried out a variety of attempts. The present methods for preparing graphene mainly include mechanical exfoliation HOPG method, redox method, silicon carbide epitaxial growth method and CVD method.

1.1.3.1 Mechanical Exfoliation Method

Mechanical exfoliation method refers to the use of tape to repeatedly stick and remove HOPG. Due to the weak van der Waals bonding force between graphite layers, the graphite could be teared apart under the sticky force of the tape. Repeated sticky can continually thin graphite, resulting in a few layers or even monolayer graphene. The method is simple to operate, but extremely inefficient and difficult to control precisely, so it's only suitable for laboratory research.

1.1.3.2 Redox Method

The basic idea of chemical oxidation reduction is similar to mechanical exfoliation, i.e. the concept of solid stripping is applied to the liquid phase. In 2006, Stankovich [33] used redox method to prepare graphene for the first time, and used the polymer to coat graphene to make it be evenly dispersed in water. Figure 1.5 shows the basic principle of the preparation of graphene by the oxidation-reduction method. First, the graphite is oxidized to form oxidized graphite, and the oxygen-containing functional group is decorated to reduce the force between the graphite layers and enhance its hydrophilicity. Then, the oxidized graphite was ultrasonically peeled off

1.1 Basic Introduction of Graphene

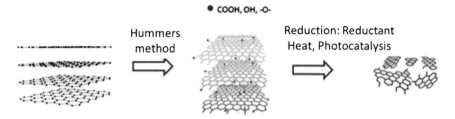

Fig. 1.5 Preparation of graphene by oxidation-reduction method. Reproduced from Ref. [34] by permission of The Royal Society of Chemistry

in water to obtain a stable oxidized graphene colloid. Finally, the reducing agent is used to reduce the graphene oxide to obtain uniformly dispersed graphene. The preparation of graphene by oxidation and reduction is one of the most promising methods to realize the macro-production of graphene.

1.1.3.3 SiC Epitaxial Growth Method

SiC epitaxial growth method also belongs to solid phase method. It refers to the use of high temperature method to decompose SiC. High temperature will make Si atoms pyrolyze and release, and then C atoms will recrystallize in the surface to form graphene. In 2006, Berger et al. [35] obtained graphene on the surface of single crystal SiC for the first time under vacuum and high temperature conditions. Deng et al. [36] studied the mechanism by which graphene epitaxially grows on the surface of monocrystalline silicon carbide, as shown in Fig. 1.6. SiC epitaxial growth method can be used to obtain high-quality graphene, but the cost is high, and the transfer of graphene is quite difficult.

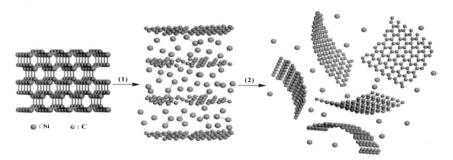

Fig. 1.6 Preparation of graphene by SiC epitaxial growth. Reproduced from Ref. [36] by permission of John Wiley & Sons Ltd.

1.1.3.4 CVD

CVD refers to the decomposition of carbon sources (methane, ethanol, etc.) under certain conditions. Then during subsequent cooling, the carbon atoms are rearranged on the substrate (copper, nickel, etc.) to form a graphene sheet, as shown in Fig. 1.7. The cost of the CVD method is low, and the size of the prepared graphene sheet is only limited by the size of the substrate. Thus, the preparation of large area continuous film can be realized by this method, and the prepared graphene can be transferred to any substrate. Therefore, it is the main method for the preparation of high quality graphene.

1.1.4 Applications of Graphene

The excellent performance described above brings graphene many potential applications in a variety of fields, as shown in Fig. 1.8. In terms of energy, its application mainly refers to solar cells [38–42]. High mobility, high transmittance, high stability, modifiability and other superior electrical properties make it possible to be used as window materials for solar cells, and can also be used as a functional layer directly involved in photoelectric conversion and other key processes. In the field of microelectronic devices, graphene has a unique and excellent carrier transport properties, making it be expected to become the basis for the next generation of integrated circuits [43, 44]. Graphene has a very high mechanical strength, so it also has potentials to be applied in microelectromechanical systems and nanoelectromechanical systems [45, 46]. High mobility and special energy band structure of graphene make it especially suitable for field effect transistors [47, 48]. From the previous description of the optics performance of graphene, it is known that graphene has good electrical conductivity, chemical stability and excellent light transmission performance. In the whole spectrum, the translucency of graphene maintained a uniform distribution, so the potential advantage of graphene in optoelectronic devices is very obvious compared to the traditional indium

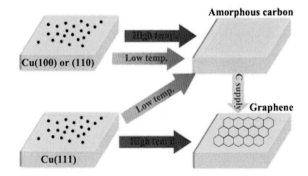

Fig. 1.7 Schematic diagram of the growth of graphene by CVD method. Reproduced from Ref. [37] by permission of The Royal Society of Chemistry (RSC) on behalf of the Centre National de la Recherche Scientifique (CNRS) and the RSC

1.1 Basic Introduction of Graphene

tin oxide transparent electrode material. Therefore, the application of graphene on the transparent electrode will become more and more prominent [49–52]. Besides, monolayer graphene has a thickness less than the DNA base sequence interval, as well as good biocompatibility and chemical stability, which has the incomparable advantage that solid-state nanopore cannot achieve in DNA sequencing [53–55]. Graphene is also one of the ideal candidate materials for flexible energy storage devices due to its high conductivity and good flexibility. For example, the use of graphene foam as a highly conductive flexible current collector can be designed and prepared to quickly charge and discharge the flexible lithium-ion battery [56, 57]. The low energy behavior of electrons in graphene is very similar to that of relativistic neutrinos, which makes it possible to pass through the obstacles encountered on the advancing path at an extremely fast rate with almost no resistance. Consequently, if the graphene-based transistor is applied in communication processing equipment, the processing speed and performance of the modulation and demodulation equipment will be greatly improved, so the graphene-based communication equipment has good development potential [58, 59]. In addition, graphene also has very good application prospects in the field of aerospace, as well as automotive batteries.

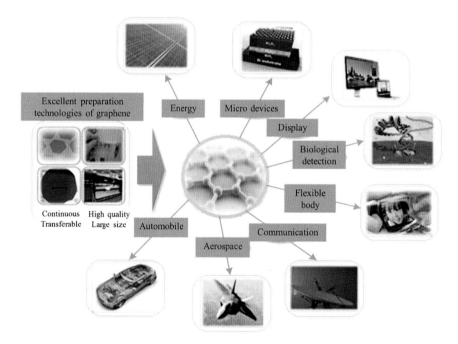

Fig. 1.8 Applications of graphene in industrial production

1.2 Processing of Graphene

At present, the preparation of large-scale high-quality graphene has been basically realized. The research hotspots have focused more on the application potential of graphene. The realization of graphene applications often involves controllable processing. For example, the nanoribbon has been shown to be able to open the zero bandgap of graphene [60], making it a nice semiconductor material with high switching ratio, and the gate voltage of the graphene can be precisely adjusted by controlling the width of the nanoribbon [38]. The processing of nanoribbons involves the tailoring of graphene. Besides, the graphene field effect transistor relates to the joining between graphene and metal electrode. The contact between metal atom and graphene sheet and the transition ratio of carrier at contact interface will have great impact on the transistor performance. The method of joining is also the main method of controlling the shape and size of the graphene structure. The graphene nanopore has been proved to have great potential application value in DNA sequencing [53], seawater desalination [61] and gas filtration [62]. To realize these applications, it is necessary to find a way to controllably fabricate graphene nanopore and modify its properties. Moreover, doping is important to realize the applications of graphene. The doping process can control the electrical and optical properties of graphene, which could enable the achievement of the application of graphene in quantum devices, solar cells.

1.2.1 Joining of Graphene

Right now, there are few reports on the joining of graphene. In 2012, Zou et al. [63] proposed the use of current joule heat to induce the connection of graphene, as shown in Fig. 1.9a. Under the action of joule heat, two pieces of overlapped graphene cannot form a connection, while butted graphene can. Then they characterized the microstructure of the joint by electron microscopy, and compared the electrical properties before and after the joining using the volt-ampere characteristic curve. In 2013, Ye et al. [64] studied the possibility of graphene joining by laser irradiation, as shown in Fig. 1.9b. It was found that under the auxiliary action of current, the laser irradiation will increase the interaction between graphene layers and convert the graphene from the monolayer Raman spectrum information to the double-layer feature. The above work is the initial exploration of graphene joining, but at present the bonding properties and the linking mechanism between graphene are still not clear, especially the joining result and influencing factors of graphene joining under different relative positions are not clear. In addition, the joining of graphene under ion beam and electron beam irradiation has not been reported.

1.2 Processing of Graphene

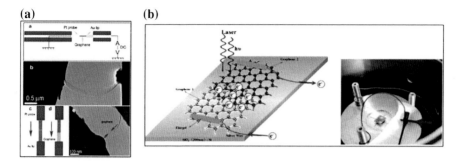

Fig. 1.9 Joining of graphene. **a** Current joule heat induced graphene joining. Reprinted from Ref. [63], Copyright 2013, with permission from Elsevier. **b** Laser irradiation induced graphene joining. Reprinted from Ref. [64], Copyright 2012, with permission from Elsevier

1.2.2 Fabrication of Graphene Nanopore

The research of graphene nanopore is now mainly focusing on how to process high quality array holes. The processing of graphene nanopore includes top-down and bottom-up methods. Among them, the top-down method is firstly preparation of single or multi-layer graphene, and then the processing of nanopore in the graphene. It mainly includes block copolymer lithography [65, 66], nano-particle lithography [67] and nano-imprint lithography [68]. The bottom-up approach refers to the etching of the substrate during graphene preparation to form a nanopore template [69–71], followed by deposition of carbon atoms to generate a graphene structure with intrinsic nanopore array. Reference [72] reviewed the main methods for the processing of graphene nanopores. Right now, the methods of preparation of graphene nanopore are aimed at batch production, and it is difficult to obtain graphene nanopores with controllable size and performance. Also, the minimum radius of nanopore is limited. Meanwhile, most of the above methods are based on mask processing, which is easy to bring external impurities pollution to affect the performance of graphene structure. The FIB method can reach a focal spot diameter of 5 nm, and the focused electron beam can get a spot diameter of 1 nm or less. The application of focused particle beam on processing of the graphene nanopore will receive more and more attention with the improvement of the equipment performance.

1.2.3 Doping of Graphene

The intrinsic graphene has a zero bandgap band structure which needs to be opened, and its application often requires controlling its conductivity. Moreover, the carbon atoms in graphene are generally bonded stably, so it is difficult make a composite with other materials. The doping modification of graphene can, on the one hand, open its zero band gap, and on the other hand, change its surface activity, to obtain controllable electrical, mechanical, and biological characteristics. The doping type of graphene is diverse, the following takes N doping as an example to introduce the doping modification method of graphene.

The nitriding of graphene mainly includes CVD [73], ammonia pyrolysis [74], arc discharge [75], and solvothermal [76]. Among them, CVD method refers to the mixing ammonia source during the CVD preparation of graphene, so that the original graphene contains nitrogen atoms. This method can generate a uniform nitrogen atoms distribution. Ammonia pyrolysis method is the heat treatment of oxide graphene in the ammonia atmosphere, so as to obtain the nitrogen-doped graphene. Arc discharge method is to generate nitrogen doped graphene through the carbon electrode arc discharge in the hydrogen and nitrogen source atmosphere. Solvent thermal method is the preparation of nitrogen-doped graphene by heat treatment in the carbon source and nitrogen source coexistent solution. As it can be seen from the above methods, all of them try to dope graphene as a whole, while it's difficult for them to dope graphene in a specified position with specified dose. Besides, usually the methods (such as the CVD method) require the use of transition metal catalysts, which can easily lead to contamination of the sample [74]. In addition, some methods are difficult to control, and the doping elements are unstable.

Usually nitrogen atoms present in three forms after being doped into graphene: graphitic-N (nitrogen atoms are bonded with three carbon atoms in the form of substitution), pyridinic-N (nitrogen atoms are bonded with two carbon atoms and form a six-membered ring with carbon atoms), pyrrolic-N (nitrogen atoms are bonded with two carbon atoms, and form a five-membered ring with carbon atoms) [73]. These three forms are also described in Fig. 1.10, in which the forms of pyridinic-N, pyrrolic-N are related to the polygonal vacancy defects in graphene.

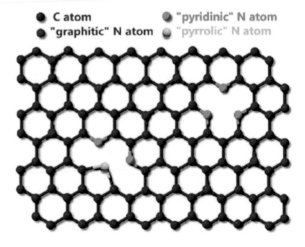

Fig. 1.10 Existence forms of nitrogen atoms in graphene structure. Reprinted with the permission from Ref. [73]. Copyright 2009 American Chemical Society

It was demonstrated that the substituted state is usually the predominant form of nitrogen atoms in graphene [77].

1.3 Particle Beam Processing and Its Application in Graphene Structure

The current method for graphene processing is mainly electrochemical method, which is difficult to achieve accurate processing in nano-scale, and difficult to be effectively controlled, and also easy to introduce impurity pollution. Particle beam processing technology is the forefront of today's manufacturing technology, so its application in the graphene processing technology has unique advantages.

1.3.1 Introduction of Particle Beam Processing

Particle beam processing technology refers to the use of laser beam, electron beam, ion beam and high pressure water jet and other high energy density beam to specially process the material or structure. It includes welding, cutting, drilling, spraying, surface modification, etching and fine processing and so on. The particle beam used in this paper refers to laser beam, electron beam, and an ion beam.

Particle beam processing technology is the forefront of the development of today's manufacturing technology, its application in graphene processing technology (graphene joining, graphene nanopore processing, graphene doping) has the following unique advantages:

1. It can achieve a high energy density, and wide range of energy density to be adjusted. The beam has excellent controlled deflection flexibility, which could ensure all-round processing, such as the graphene joining under different graphene relative positions.
2. The particle beam can also be focused into a very fine beam to reach the focal spot of micrometer (laser) or even nanoscale (electron beam and ion beam), so that the graphene nanoribbon and nanopore can be finely processed. Also, it is possible to join and dope graphene at specified location.
3. The real-time monitoring technology in the process is becoming more and more mature, making it possible to finely control the process of graphene. Such as precisely controlling the dose of ion doping, and precisely setting the energy of the particle beam for graphene joining.
4. Particle beam processing of graphene involves the collision and charge transfer between atoms, the quantum effect of nanometer scale and the transient effect of ultrashort time. The edge of graphene may produce plasmon excitation, and the graphene structure may produce Coulomb explosion. These phenomena determine the specificity of the particle beam processing of graphene nanostructures.

1.3.2 Research Status of Particle Beam Processing Graphene

The study of graphene joining by particle beam irradiation is rarely reported. Since 2001, scholars have proposed the use of particle beam irradiation method to join CNTs [78–80]. Figure 1.11 shows the results of joining the CNTs using particle beam irradiation. It can be seen that, on the one hand, the irradiation of the particles can form the connection between two CNTs. This connection is due to the formation of new chemical bonds between the different CNTs. The particle beam irradiation parameters can have large effect on the performance of the joint. On the other hand, particle beam irradiation also brings a certain degree of amorphization to the irradiation region, leading to the occurrence of vacancy defects at the joint. These vacancy defects will affect the mechanical properties of the nanotube joint (Fig. 1.12). Because graphene and CNTs are allotrope composed of carbon element, they may have some similar joining mechanisms under the irradiation of particle beam. The joining of CNTs can provide some inspiration for graphene joining. However, graphene has two-dimensional planar structure characteristics, so that it has different joining phenomenon from the CNTs, which needs to be further analyzed. At the same time, particle beam irradiation will also bring collision defects to the joint, evaluation of the impact of these defects on its mechanical and electrical properties is extremely important to the application of graphene joint. Therefore, it needs to conduct further study for joining of graphene by particle beam irradiation and evaluation of the joint performance.

In 2013, Bangert et al. [81] firstly proposed the use of low-energy ion implantation method for graphene doping with B and N elements, for which the doping diagram is shown in Fig. 1.13a. The doping results were analyzed by high resolution TEM. It was found that when the energy was low, the implanted ion beam was mainly present in the graphene structure as substituted state, and a small amount of ions were doped accompanied by the formation of the defective state (as

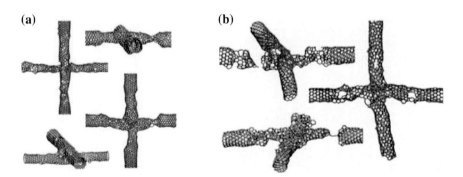

Fig. 1.11 Joining of CNTs by particle beam irradiation. **a** The results of the joining of CNTs; **b** Influence of parameters. Reprinted with permission from Ref. [79]. Copyright 2002 by American Physical Society

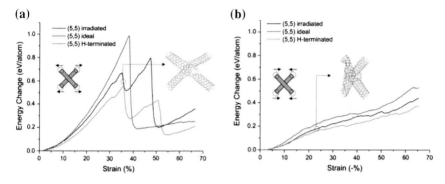

Fig. 1.12 **a** Tensile and **b** compressive mechanical properties of the CNTs joined by particle beam irradiation. Reprinted with the permission from Ref. [80]. Copyright 2004 American Chemical Society

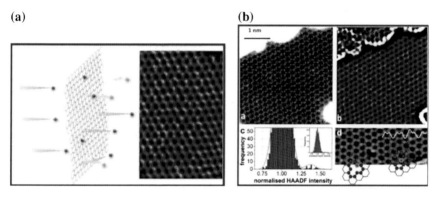

Fig. 1.13 Graphene doping by low-energy ion implantation. **a** Illustration of the doping process. **b** Nitrogen ion doped results. Reprinted with the permission from Ref. [81]. Copyright 2013 American Chemical Society

shown in Fig. 1.13b). Subsequently, Xu et al. [82] further analyzed the doping results of N, B and F elements for graphene by low-energy ion implantation using XPS and Raman spectroscopy. It was found that low energy ion implantation doping would damage the graphene structure and the degree of damage increased with the increase of energy and dose of the ion beam. Ion implantation can overcome the weak chemical activity of graphene surface, so it has important application value in graphene doping. The above studies confirm the feasibility of graphene doping by low-energy ion implantation method, but due to the limitations of the current experimental technologies, the dynamics mechanism of atomization of graphene doping and the destruction of graphene during the ion irradiation process are not that clear, and a detailed discussion of the influence of incident ion parameters (energy and dose) on the doping process is lacking. In addition, the incident low energy ions collide with the carbon atoms in the graphene structure, which changes the mechanical and electrical properties of the graphene, and these changes in performance may have an important effect on the applications of graphene. Therefore, it is necessary to further analyze the properties of doped graphene and discuss the potential application value.

The focused electron beam has been shown to be useful for processing nanoscale holes in graphene [83–85]. Stable nanopores can be obtained by controlling the number of layers, temperature and other factors of graphene, and the size of nanopores can be reduced to be within 5 nm. Electron beam processing of nano-holes need a high acceleration voltage, so the equipment requirements are higher; ion beam has a greater weight, theoretically ion beam can be more effective in nanopore processing. However, the current focus radius of the ion beam is limited, so the resulting nanopore size can only reach tens of nanometers. The accuracy of ion beam processing can be further improved by using the mask version. At the same time, the focus radius of ion beam itself is also decreasing with the development of equipment. On the one hand, because the transform mechanism of the collision energy between the FIB and the electron beam with the carbon atoms in graphene is different, the mechanisms and optimal parameters for them to be used in the processing of nanopores are different, which needs to be discussed in depth. On the other hand, the performance and dynamics mechanism of nanopore processing for the substrate supported and suspended graphene are also different, which need to be figured out. In addition, in order to reduce the influence of redundant parameters on the performance of equipment and nanopore, it is necessary to discuss the effect of ion beam and electron beam parameters on nanopore processing. For the obtained nanopore structure, it is necessary to withstand great mechanical stress in the applications of gene detection and brine purification [86], and nanopore can also significantly improve the switching performance of graphene sensors based on its electrical properties [87]. Therefore, it is also very important to explore the properties of obtained graphene nanopore.

1.4 Problem Introduction

As a two-dimensional new material, graphene shows special optical, electrical, thermal and mechanical properties due to its sp^2 hybrid hexagonal honeycomb lattice and single atom thickness. These properties bring great potential applications of graphene in society, such as solar cells, transistors, touch screen, biosensors and so on. At present, the research hotspot of graphene is focused on the application of graphene, and the application of nanomaterials cannot be separated from its nanostructure processing. Processing can change the structural morphology of graphene and even control its properties to obtain the required graphene structure, and to promote the application of graphene.

The controllable processing of graphene consists of three aspects: on the one hand, the need to join multiple pieces of graphene to effectively control the size and

1.4 Problem Introduction

shape of graphene. On the other hand, it is necessary to dope the graphene structure to open the energy band, and control its electrical properties. In addition, it is needful to process graphene nanopore for molecular sequencing, seawater purification, gas molecular sieves and so on. The current processing methods are mainly concentrated in a number of physical and chemical methods, such as joining of graphene by current joule heat, photoetching of nanopore using mask template, doping of graphene in ammonia source atmosphere. These methods have varying degrees of shortages, such as the complex operation, high demand for equipment, limited processing accuracy, easy to bring pollution, unable to be doped in specified region and so on.

Particle beam processing technology has the characteristics of high precision, high efficiency, mature control method and fast development speed. It has obvious advantages in the processing of graphene nanostructures: such as in the joining of graphene, it can control the particle beam species to achieve the different forms of joint, and the performance of joint can also be controlled by changing the particle beam parameters. For the graphene doping, low-energy ion implantation can be used to dope graphene at the specified location with specified concentration. For the processing of graphene nanopore, focused electron beams and ion beams can be used to get high precision nanopore structures. The current research focuses on the interaction mechanism between the particle beam and graphene. As for the joining of graphene by particle beam irradiation, it's rarely reported, and there are few studies on the doping of graphene by particle beam implantation. Meanwhile, the discussion about the influence of particle beam irradiation parameters on the fabrication of graphene nanopore is limited. Besides, the current researches on the mechanism of graphene doping, joining and nanopore processing, and the properties of the obtained nanostructures are scarce.

Based on this, the processing of nanostructures of graphene by particle beam irradiation was studied in this paper. Firstly, the phenomena of graphene structure irradiated by particles with different energy were studied experimentally. The mechanism of the interaction between carbon atoms in graphene and irradiated particles was investigated by atomic simulation method. Then, low-energy ion implantation was used to dope graphene, for which experimental methods were conducted to explore the possibility of this doping method, and computational simulation method was used to uncover the doping mechanism. After that, FIB and laser beam were proposed to join graphene, and the bonding mechanisms between different graphene films were discussed according to different types of particle beams. Next, the mechanism and principle of the graphene nanopore processing by ion beam and electron beam irradiation with high energy were investigated by taking substrate supported and suspended graphene into consideration. Finally, based on the phenomena and mechanisms of the graphene nanostructure processing, the mechanical and electronic transport properties of the obtained graphene nanostructures were further discussed, and the application potential of graphene nanostructures was analyzed.

1.5 Research Content

In this paper, we discussed the scientific problems in the processing of graphene nanostructures by particle beam irradiation. Starting from the interaction mechanism between different types and different energy of particle beams and graphene, the experimental possibility for the doping of graphene by low energy particle beam implantation, joining of graphene by moderate energy particle beam irradiation and fabrication of graphene nanopore by high energy particle beams was explored. And then the mechanisms of graphene nanostructures processing were analyzed by atomic simulations. Finally, the mechanical and electrical properties of the nanostructures were analyzed, which laid the foundation for the realization of the applications of graphene.

The main contents of this research are as follows:

Interaction mechanism between the particle beams and graphene

1. Morphological changes and structural damage of graphene under ultrafast laser action
2. Effects of different energy ion beams on graphene and the influence of substrate
3. The mechanism of the change of graphene structure under different energy electron beam irradiation

Doping of graphene by low energy ion implantation and the properties of the doped structures

1. Experimental study of graphene doing by low energy nitrogen ion beam and the mechanism
2. Analysis of the mechanical properties and electronic transport properties of doped graphene

Joining of graphene by particle beam irradiation and joint performance analysis

1. Experimental phenomenon and mechanism analysis of laser and ion beam induced graphene joining
2. Analysis of mechanical properties and electron transport properties of graphene joint

Fabrication of graphene nanopore by particle beam irradiation and the nanopore performance analysis

1. Experimental and theoretical analysis of graphene nanopore processing by FIB and electron beam
2. Mechanical properties and electronic transport properties analysis of graphene nanopore.

References

1. Chen YS, Huang Y et al (2013) Graphene: new type of two-dimensional carbon nanomaterials. Science Press, Beijing (in Chinese)
2. Kroto HW, Health JR, O'Brien SC et al (1985) C60: Buckminsterfullerene. Nature 318: 162–163
3. Iijima S (1991) Helical microtubules of graphitic carbon. Nature 354:56–58
4. Mermin ND (1968) Crystalline order in two dimensions. Phys Rev 176:250–254
5. Novoselov K, Geim AK, Morozov SV et al (2004) Electric field effect in atomically thin carbon films. Science 306:666–669
6. Boehm HP, Setton R, Stumpp E (1986) Nomenclature and terminology of graphite intercalation compounds. Carbon 24:241–245
7. Boehm HP, Setton R, Stumpp E (1994) Nomenclature and terminology of graphite intercalation compounds. Pure Appl Chem 66:1893–1901
8. Fitzer E, Kochling KH, Boehm HP et al (1995) Recommended terminology for the description of carbon as a solid. Pure Appl Chem 67:473–506
9. McNaught AD, Wilkinson A (1997) IUPAC in compendium of chemical terminology, 2nd edn. Blackwell Scientific, Oxford
10. China graphene industry technology innovation strategic alliance, terms and definitions of graphene materials (2014) Q/LM01CGS001–2013 (in Chinese)
11. Geim AK, Novoselov KS (2007) The rise of graphene. Nat Mater 6:183–191
12. Zhu HW, Xu ZP, Xie D et al (2011) Graphene-the structure, preparation methods and properties characterization. Tsinghua University Press, Beijing (in Chinese)
13. Hass J, de Heer WA, Conrad EH (2008) The growth and morphology of epitaxial multilayer graphene. J Phys Condens Mater 20:323202
14. Meyer JC, Geim AK, Katsnelson MI et al (2007) The structure of suspended graphene sheets. Nature 446:60–63
15. Lee C, Wei X, Kysar JW et al (2008) Measurement of the elastic properties and intrinsic strength of monolayer graphene. Science 321:385
16. Frank IW, Tanenbaum DM (2007) Mechanical properties of suspended graphene sheets. J Vac Sci Technol B 25:2558–2561
17. Huang X, Qi X, Boey F et al (2012) Graphene-based composites. Chem Soc Rev 41:666–686
18. Bolotin KI, Sikes KJ, Jiang Z et al (2008) Ultrahigh electron mobility in suspended graphene. Solid State Commun 146:351–355
19. Kim KS, Zhao Y, Jang H et al (2009) Large-scale pattern growth of graphene films for stretchable transparent electrodes. Nature 457:706–710
20. Zhang YB, Tan YW, Stormer HL et al (2005) Experimental observation of the quantum Hall effect and Berry's phase in graphene. Nature 438:201
21. Novoselov KS, Jiang D, Schedin F et al (2005) Two-dimensional atomic crystals. Proc Natl Acad Sci U S A 102:10451
22. Elias DC, Nair RR, Mohiuddin T et al (2009) Control of graphene's properties by reversible hydrogenation: evidence for graphene. Science 323:610–613
23. Castro N, Guinea F, Peres N et al (2009) The electronic properties of graphene. Rev Mod Phys 81:109
24. Nair RR, Blake P, Grigorenko AN et al (2008) Fine structure constant defines visual transparency of graphene. Science 320:1308
25. Wang F, Zhang Y, Tian C et al (2008) Gate-variable optical transitions in graphene. Science 320:206
26. Bonaccorso F, Sun Z, Hasan T et al (2010) Graphene photonics and optoelectronics. Nat Photonics 4:611–612
27. Hendry E, Hale PJ, Moger J et al (2010) Coherent nonlinear optical response of graphene. Phys Rev Lett 105:097401

28. Balandin AA (2011) Thermal properties of graphene and nanostructured carbon materials. Nat Mater 10:569–581
29. Balandin AA, Ghosh S, Bao WZ et al (2008) Superior thermal conductivity of single-layer graphene. Nano Lett 8:902–907
30. Zuo X, He S, Li D et al (2010) Graphene oxide-facilitated electron transfer of metalloproteins at electrode surfaces. Langmuir 26:1936–1939
31. Stoller MD, Park SJ, Zhu YW et al (2008) Graphene-based ultracapacitors. Nano Lett 8:3498–3502
32. Schniepp HC, Li JL, McAllister MJ et al (2006) Functionalized single graphene sheets derive from splitting graphite oxide. J Phys Chem B 110:8535–8539
33. Stankovich S, Piner RD, Chen XQ et al (2006) Stable aqueous dispersions of graphitic nanoplatelets via the reduction of exfoliated graphite oxide in the presence of poly(sodium 4-styrenesulfonate). J Mater Chem 16:155–158
34. Concha BN, Eugenio C, Carlos MG et al (2012) Influence of the pH on the synthesis of reduced graphene oxide under hydrothermal conditions. Nanoscale 4:3977–3982
35. Berger C, Song Z, Li T et al (2004) Ultrathin epitaxial graphite: 2D electron gas properties and a route toward graphene-based nanoelectronics. J Phys Chem B 108:19912–19916
36. Deng D, Pan X, Zhang H et al (2010) Freestanding graphene by thermal splitting of silicon carbide granules. Adv Mater 22:2168–2171
37. Hu B, Ago H, Orofeo CM et al (2012) On the nucleation of graphene by chemical vapor deposition. New J Chem 36:73–77
38. Li XL, Wang XR, Zhang L et al (2008) Chemically derived ultrasmooth graphene nanoribbon semiconductors. Science 319:1229–1232
39. Wang X, Zhi LJ, Müllen K (2008) Transparent, conductive graphene electrodes for dye-sensitized solar cells. Nano Lett 8:323–327
40. Miao XC, Tongay S, Petterson MK et al (2012) High efficiency graphene solar cells by chemical doping. Nano Lett 12:2745–2750
41. Wang JT, Ball JM, Barea EM et al (2014) Low-temperature proceed electron collection layers of graphene/TiO$_2$ nanocomposites in thin film perovskite solar cells. Nano Lett 14:724–730
42. Li XM, Zhu HW, Wang KL et al (2010) Graphene-on-silicon Schottky junction solar cells. Adv Mater 22:2743–2748
43. Han SJ, Garcia AV, Oida S et al (2013) Graphene radio frequency receiver integrated circuit. Nat Commun 5:3086
44. Lin YM, Garcia AV, Han SJ et al (2011) Wafer-scale graphene integrated circuit. Science 332:1294–1297
45. Bunch JS, van der Zande AM, Verbridge SS et al (2007) Electromechanical resonators from graphene sheets. Science 315:490–493
46. Smith AD, Niklaus F, Paussa A et al (2013) Electromechanical piezoresistive sensing in suspended graphene membranes. Nano Lett 13:3237–3242
47. Rojas FM, Rossier JF, Brey L et al (2008) Performance limits of graphene-ribbon field-effect transistors. Phys Rev B 77:045301
48. Farmer DB, Chiu HY, Lin YM et al (2009) Utilization of a buffered dielectric to achieve high field-effect carrier mobility in graphene transistors. Nano Lett 9:4474–4478
49. Blake P, Brimicombe PD, Nair RR et al (2010) Doped graphene electrodes for organic solar cells. Nanotechnology 21:505204
50. Li X, Zhu Y, Cai W et al (2009) Transfer of large-area graphene films for high-performance transparent conductive electrodes. Nano Lett 9:4359–4363
51. Bae S, Kim H, Lee Y et al (2010) Roll-to-roll production of 30-inch graphene films for transparent electrodes. Nat Nanotech 5:574–578
52. Wu J, Becerril Bao Z et al (2008) Organic solar cells with solution-processed graphene transparent electrodes. Appl Phys Lett 92:263302
53. Merchant CA, Healy K, Wanunu M et al (2010) DNA translocation through graphene nanopores. Nano Lett 10:2915–2921

References

54. Avdoshenko SM, Nozaki D, da Rocha CG et al (2013) Dynamics and electronic transport properties of DNA translocation through graphene nanopores. Nano Lett 13:1969–1976
55. Sathe C, Zou X, Leburton JP et al (2011) Computational investigation of DNA detection using graphene nanopores. ACS Nano 5:8842–8851
56. Chen ZP, Ren W, Gao L et al (2011) Three-dimensional flexible and conductive interconnected graphene networks grown by chemical vapour deposition. Nat Mater 10:424–428
57. Wang DW, Li F, Zhao J et al (2009) Fabrication of graphene/polyaniline composite paper via in situ anodic electropolymerization for high-performance flexible electrode. ACS Nano 3:1745–1752
58. Palacios T, Hsu A, Wang H (2010) Applications of graphene devices in RF communications. IEEE Commun Mag 48:122–128
59. Abadal S, Alarcón E, Lemme M et al (2013) Graphene-enabled wireless communication for massive multicore architectures. IEEE Commun Mag 51:137–143
60. Son YW, Cohen ML, Louie SG (2006) Energy gaps in graphene nanoribbons. Phys Rev Lett 97:216803
61. Tanugi DC, Grossman JC (2012) Water desalination across nanoporous graphene. Nano Lett 12:3602–3608
62. Sint K, Wang B, Král P (2008) Selective ion passage through functionalized graphene nanopores. J Am Chem Soc 130:16448–16449
63. Zou R, Zhang Z, Xu K (2012) A method for joining individual graphene sheets. Carbon 50:4965–4972
64. Ye X, Huang T, Lin Z et al (2013) Lap joining of graphene flakes by current-assisted CO_2 laser irradiation. Carbon 61:329–335
65. Kim BH, Kim JY, Jeong SJ et al (2010) Surface energy modification by spin-cast, large-area graphene film for block copolymer lithography. ACS Nano 4:5464–5470
66. Bai JW, Cheng R, Xiu F et al (2010) Very large magnetoresistance in graphene nanoribbons. Nat Nanotechnol 5:655–659
67. Sinitskii A, Tour JM (2010) Patterning graphene through the self-assembled templates: toward periodic two-dimensional graphene nanostructures with semiconductor properties. J Am Chem Soc 132:14730–14732
68. Liang XG, Jung YS, Wu S et al (2010) Formation of bandgap and subbands in graphene nanomeshes with sub-10 nm ribbon width fabricated via nanoimprint lithography. Nano Lett 10:2454–2460
69. Ning GQ, Fan Z, Wang G et al (2011) Gram-scale synthesis of nanomesh grahene with high surface area and its applications in supercapacitor electrodes. Chem Commun 47:5976–5978
70. Wang M, Fu L, Gan L et al (2013) CVD growth of large area smooth-edged graphene nanomesh by nanosphere lithography. Sci Rep 3:1238
71. Safron N, Kim M, Gopalan P et al (2012) Barrier-guided growth of micro- and nano-structured graphene. Adv Mater 24:1041–1045
72. Yuan WJ, Chen J, Shi G (2014) Nanoporous graphene materials. Mater Today 17:77–85
73. Wei D, Liu Y, Wang Y et al (2009) Synthesis of N-doped graphene by chemical vapor deposition and its electrical properties. Nano Lett 9:1752–1758
74. Sheng ZH, Tao L, Chen JJ et al (2011) Catalyst-free synthesis of nitrogen-doped graphene via thermal annealing graphite oxide with melamine and its excellent electrocatalysis. ACS Nano 5:4350–4358
75. Suezawa MS, Sumino KJ, Harada HF et al (1986) Nitrogen oxygen complexes as shallow donors in silicon crystals. J Appl Phys 25:859–861
76. Deng DH, Pan X, Yu L et al (2011) Toward N-doped graphene via solvothermal synthesis. Chem Mater 23:1188–1193
77. Iwazaki T, Obinata R, Sugimoto W (2009) High oxygen-reduction activity of silk-derived activated carbon. Electrochem Commun 11:376–378
78. Terrones M, Banhart F, Grobert N et al (2002) Molecular junctions by joining single-walled carbon nanotubes. Phys Rev Lett 89:075505

79. Krasheninnkov AV, Nordlund K, Keinonen J (2002) Ion-irradiation-induced welding of carbon nanotubes. Phys Rev B 66:245403
80. Jang I, Sinnott SB (2004) Molecular dynamics simulation study of carbon nanotube welding under electron beam irradiation. Nano Lett 4:109–114
81. Bangert U, Pierce W, Kepaptsoglou DM et al (2013) Ion implantation of graphene-toward IC compatible technology. Nano Lett 13:4902–4907
82. Xu Y, Zhang K, Brüsewitz C et al (2013) Investigate of the effect of low energy ion beam irradiation on mono-layer graphene. AIP Adv 3:072120
83. Xu T, Xie X, Sun L (2013) Fabrication of nanopores using electron beam. Paper presented at NEMS2013, Suzhou, China, 7–10 Apr 2013
84. He K, Robertson AW, Gong C et al (2015) Controlled formation of closed-edge nanopores in graphene. Nanoscale 7:11602
85. Lu N, Wang J, Floresca HC et al (2012) In situ studies on the shrinkage and expansion of graphene nanopores under electron beam irradiation at temperatures in the range of 400–1200 °C. Carbon 50:2961–2965
86. Siwy ZS, Davenport M (2010) Nanopores: GRAPHENE opens up to DNA. Nat Nanotech 5:697–698
87. Bai J, Zhong X, Jiang S et al (2010) Graphene nanomesh. Nat Nanotechnol 5:190–194

Chapter 2
Experiment Approaches and Simulation Methods

2.1 Synthesis and Characterization of Graphene Specimen

In this section, firstly the preparation methods of graphene samples were introduced, including the synthesis of monolayer and multilayered graphene by CVD, substrate etching and graphene transfer. Secondly, the main characterization methods of graphene before and after processing were introduced. Finally, the equipment for processing of graphene by particle beam irradiation was introduced.

2.1.1 Preparation of Monolayer and Multilayer Graphene Specimens

2.1.1.1 CVD Preparation Method

The graphene was synthesized by CVD method. For the preparation of multi-layer graphene, the name and model of the equipment used in this paper is tube resistance furnace GERO F100-750/13 with a furnace size 350 × 950 × 420 mm, and furnace cavity diameter Ø100 mm. The maximum temperature can reach 1300 °C. Figure 2.1 shows a schematic diagram of the CVD setup for the preparation of graphene.

The CVD method for the preparation of graphene films mainly involves the following steps:

(1) Cleaning of the heating pipe. Use a volatile liquid (alcohol) to clean the pipe in the tube.
(2) Selection of growth substrate. In order to grow monolayer and a few layers of graphene samples, the substrates on which graphene grows were selected as Pt and Cu, respectively. The growth of graphene on the Pt substrate depends on the dissolution of the carbon atoms in the carbon atmosphere at high

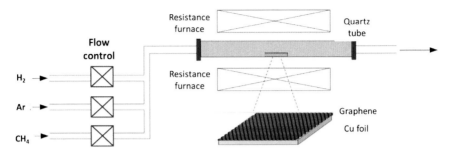

Fig. 2.1 Schematic diagram of the CVD apparatus for the preparation of graphene

temperature and crystallization at low temperature. The Cu-based growth graphene is directly crystallized on the surface by the deposited carbon atoms. Due to the limitation of the growth environment, the few layers graphene was grown by Cu substrate, and the monolayer graphene film was grown by Pt substrate. Most of the experiments used few layers graphene samples. Therefore, only the growth of graphene on Cu substrate was introduced in detail. The copper foil was manufactured by Alfa Aesar and has a mass purity of 99.8% and a thickness of 25 μm. Before the experiment, the copper foil was pre-cleaned by ultrasonic and cut into the appropriate size.

(3) Sealing and pumping the protective gas. The copper foil was placed at constant temperature area of the tube furnace, and then sealed tube furnace. Then the argon was pumped in, and sealing results were checked to strictly guarantee the gas seal.

(4) Hydrogen input. Set the program for heating with a heating rate of 10 °C/s. When the furnace temperature rose to 400 °C, the hydrogen could be pumped in with a gas flow as shown in Table 2.1.

(5) Carbon source input. When the temperature rose to 1000 °C, the gas flow of argon and hydrogen was adjusted, and this temperature was kept for 60 min. Afterwards, carbon source (methane) began to be inputted in.

(6) High temperature reaction. Methane gas began to decompose under the reduction of hydrogen for about 10 min. The furnace temperature during the reaction was 1000 °C. After the completion of the reaction, we stopped the methane, and moved the copper foil to low temperature area of the furnace.

(7) Sample cooling. Turn off the hydrogen and resistance furnace power supply so that the sample was cooled to room temperature with argon.

Table 2.1 Gas flow at different stages of graphene preparation by CVD method

Gas	Heating (mL/min)	Growth (mL/min)	Air cooling (mL/min)	Final (mL/min)
Ar	0.7	1.6	0.5	0
H_2	400	400	0	0
CH_4	0	30	0	0

2.1 Synthesis and Characterization of Graphene Specimen

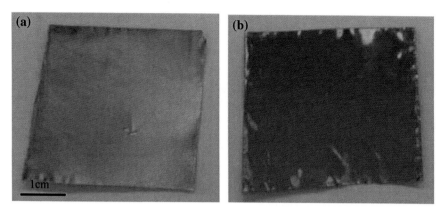

Fig. 2.2 Appearance of copper substrate **a** before and **b** after growth of graphene

(8) Samples taken out. Remove the sample from the furnace and store it in a suitable environment. Then the laboratory equipment was recovered.

The final graphene sample prepared on the copper substrate is shown in Fig. 2.2. The copper base is bright yellow before growth of graphene and becomes dark yellow after growth of graphene.

2.1.1.2 Substrate Etching and Graphene Transfer

The transfer of the substrate mainly involves the etching of the growth substrate and the reloading of the target substrate. For the graphene grown on Pt substrate, the surface PMMA was used to help with the etching. PMMA can ensure the quality of monolayer graphene film, but the subsequent removal process is complicated. For the graphene grown on the copper foil substrate, the $FeCl_3$ solution was directly used to etch the substrate. Because the strength of graphene film is high, it can be transferred to the target substrate quickly and easily.

The transfer of graphene was divided into the following steps: (1) cutting the wafer and cleaning. The silicon wafer used for the experiment is a 4-in. single-side polished monocrystalline silicon with 300 nm thickness oxide layer. The thickness of the wafer is 360 μm. (2) Preparation of the etchant. Corrosive solution was made up by the mixed solution of 0.5 mol/L ferric chloride and 0.5 mol/L hydrochloric acid. (3) Etching of copper substrate. The copper-supported graphene was placed in the culture dish for etching for about 60 min. (4) Cleaning by deionized water. After the completion of the etching, the floating graphene was taken out and placed in deionized water for about 60 min, and the process was repeated 3–4 times until the graphene was perfectly cleaned. (5) Transferring to the target substrate. The graphene film was taken out from deionized water by using the silicon wafer. At this time, the graphene was transferred from the initial substrate to the target substrate. (6) Characterization of the samples. The samples were characterized with a microscope to evaluate the quality. Figure 2.3 shows the transferred graphene on the silicon substrate.

2.1.2 The Main Characterization Methods of Graphene Sample

The characterization of the structural morphology and performance of graphene samples before and after processed by particle beam irradiation needs a variety of modern microscopic analysis techniques to be used. Different analytical methods can be used to characterize graphene from different perspectives. The main characterization methods used in this paper are presented below.

(1) Optical microscopy

Optical microscope is a preliminary method of characterizing the microstructure of the material. Due to the substrate and the graphene structure reflect differently to the incident light, the graphene film on the silica substrate has a special color, which makes the structure of the graphene be resolvable and distinguishable. In this paper, the model of the optical microscope equipment is OLYMPUS BX51M, The amplification ranges from 50 to 1000 times.

(2) Scanning electron microscopy

SEM can be used to observe the microstructure of graphene, the surface morphology of the substrate, and the structural changes before and after graphene processing. In this paper, the SEM device is LEO-1530, with a resolution of 1 nm, and magnification of 20–900,000 times. The accelerating voltage of the electron beam is 100 V–30 kV.

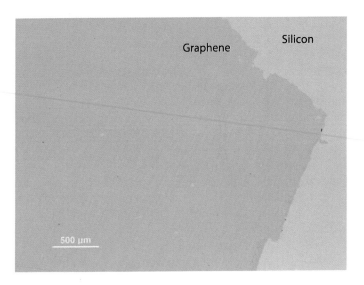

Fig. 2.3 Graphene on a silicon substrate

2.1 Synthesis and Characterization of Graphene Specimen

(3) High resolution transmission electron microscopy

TEM was used to process the graphene nanopore structure. The structure information of graphene and the number of layers of graphene and graphene edge morphology can also be observed by TEM. The model is Tecnai G2 F20 S-Twin. The maximum acceleration voltage is 200 kV with a point resolution of 0.24 nm, and the minimum beam spot size is 0.3 nm.

(4) Raman spectroscopy

Raman spectroscope is widely used in the characterization of carbon materials [1]. The graphene usually has three main characteristic peaks, namely, the D peak caused by defects near 1350 cm^{-1}, the G peak caused by E_{2g} vibration near 1584 cm^{-1}, and the overtone peak 2D peak near 2680 cm^{-1}. D peak represents the defects in the structure of graphene, which can reflect the damage of graphene structure before and after processing in graphene doping and nanopore fabrication. The peak intensity ratio and position information of G peak and 2D peak can reflect the layer information of graphene. Usually, the 2D peak of single layer graphene is stronger than G peak. If the overlapped graphene is joined with each other, the intensity of 2D peak will decrease gradually when the number of layers increases, and the location will shift right as well. Therefore, the joining between graphene layers can be determined by analyzing the peak intensity ratio of the G peak and the 2D peak. The mode of Raman analysis equipment is LabRAM HR Evolution. The laser spot used was 1.25 μm, and the analysis band was 1000–3500 cm^{-1}. The acquisition time was 20 s, and the power of the laser was 0.38 mW. In addition, Raman Imaging was used to analyze the changes of graphene structure in the processing area in batch, and the uniformity of the results was also analyzed.

(5) Atomic force microscopy

AFM can analyze the surface fluctuation of graphene from the atomic level, and the changes of graphene structure before and after irradiation by particle beam. Through the contact mode, the height information of graphene at the boundary before and after joining can be analyzed, which can be used to decide whether joint is formed or not. In addition, by analyzing the morphological information of the nanopore structure, it is possible to describe the size, depth and three-dimensional structure information. The device model is Veeco USA.

(6) X-ray photoelectron spectroscopy

The XPS can analyze the composition and structure of the surface elements of the graphene films. The doping information of the graphene structure and the chemical bonding formed between the elements can be obtained by the analysis of the peak fitting of the chemical elements. XPS analysis equipment model is ESCALA8 250 Xi, and the best energy resolution is less than 0.45 eV.

2.1.3 The Main Experimental Equipment for Graphene Processing

1. SEM/FIB double beam system

For the processing of nanopore and nanoribbon in supported single layer graphene, a SEM/FIB double beam system was used in this paper. Among them, the FIB was used to process the graphene nanostructures, and the SEM was used to observe the processing in situ. The focal ion source used in the apparatus was gallium ion, and the ion beam energy was fixed at 30 keV. The ion beam was focused into a beam spot with a diameter of about 5 nm using a focusing system. By selecting different ion beam streams, and controlling the moving speed of the workbench, different levels of micro-etching can be achieved for the sample. SEM used the field emission filament. In order to reduce the destruction of the electron beam on the sample, an electron beam acceleration voltage of 5 kV was adopted, and the electronic current was 1 pA. The equipment was manufactured by TESCAN (Czech), and model is LYRA 3 FEG.

2. Low energy ion implantation machine

The experimental equipment used for low energy ion implantation is a plasma immersion ion implantation machine, which utilizes an inductive coupling to produce a high-density plasma in a reaction chamber by a set of high-power radio frequency excitation power sources. In the meantime, another group of pulsed bias power sources directs ions perpendicularly move to the sample, which could inject the ions into the interior of the sample. This system can realize the purpose of high dose low energy injection with low damage. The gas flow rate of the ion implantation process was 250 mL/min, and the power was set as 800 w, with a pulse width of 30 μs. The energy of the implanted ions and the injection time were controlled to realize the injection of ions with different energy and dose. The excitation power was 13.56 MHz and 1500 W, with an injection bias of 0–5 kV. The equipment model is PIII-200.

3. Femtosecond laser system

The laser-processing of micron structure on graphene used COHERENT Company's semiconductor pump-mode injection molding titanium sapphire laser amplifier. The laser center wavelength was 800 nm, and the pulse width was 50 fs, with a maximum repetition frequency of 1000 Hz, and the focused spot of the laser after focusing through the lens was about 10 μm. The initial output power of the femtosecond laser was 3.52 W, and the laser power was controlled by two-stage attenuation method, so as to study the processing damage of the graphene structure under different energy laser.

2.2 Introduction of MD Simulation

2.2.1 Concepts

MD simulation is an important method in computational materials science. MD mainly refers to that according to the initial conditions of the system in the ground state, the use of computer simulations to integrally solve the motion state of the system under the external conditions, based on the empirical or semi-empirical parameters. Through averaging of the molecular microscopic state in the time and space, we can get the macro physical quantities of the system, such as the system temperature, pressure, energy and so on. It can also get the structural information of the system in equilibrium state. MD simulation is divided into the first-principle MD simulation based on quantum mechanics and classical MD simulation based on Newton's second law. Among them, the classic MD can solve a system with tens of thousands of atoms on an ordinary computer, and it can also investigate the phenomenon and mechanism of graphene processed by particle beam irradiation from atomic scale level. In this article, if it is not specifically described, the MD particularly refers to the classical MD simulation.

2.2.2 Basic Principles of Classic MD

Classical MD refers to the using of computer simulation to get the structure and properties of a system made up of nuclei and nuclear electrons. The movement of each nucleus follows Newton's law of motion and the nucleus is considered to move under the average potential of all other nuclei and electrons. For the interaction of nuclei-nuclei, nuclei-electron, and electron-electron, analytical potential is adopted. Classic MD is generally divided into the following steps:

1. Determination of the research model. Extracting the corresponding physical model according to the problem to be studied, and then establishing the corresponding MD model.
2. Selection of the force field. Selecting the appropriate force field to describe the interatomic interaction potential energy and determining the parameters of the force field.
3. Boundary conditions. The supercell of MD simulation is only a repeating unit of the macro system. In order to simulate the free system or periodic system, it is needed to set a fixed, shrink wrap or periodic boundary conditions.
4. System initialization. Calculating the minimum energy of the system with a given initial conditions to get the steady-state structure of the system under initial equilibrium conditions, which could provide the equilibrium structure for the following dynamics simulation under the external conditions.

5. Solution of Newton equation. Under the external conditions, the force and potential energy of the particles in the system are solved and the iterations are repeated until the requirements are satisfied.
6. Output of the results, averaging of macro physical quantity.

It can be seen from the simulation process that the selection of the interaction potential between atoms is very important for the simulation results of graphene materials. The choice of numerical integration algorithm, the simulation system ensemble and the results statistical method have great influence on the accuracy of the results. They will be described in detail in next section.

2.2.3 Atomic Interaction Force

The key to the classical MD simulation lies in the selection of the potential energy function for the interaction between atoms. Because of the complexity of the simulation system and the different interactions between different elements, it is difficult to describe the interaction between different simulation systems and different elements by using a unified potential energy function. In the actual simulation process, it is often necessary to select the appropriate potential function according to the change of the simulation system and the composition of the atomic structure. In this paper, AIREBO [2] was used to describe the interaction between graphene carbon atoms, and L-J potential was used to describe the interaction between carbon atoms in graphene and Si, O atoms in substrate. ZBL [3] potential was selected to describe the collision between the incident ions and the target atoms, and Tersoff [4] potential was adapted to describe the interaction between N, B and C atoms, and also the interaction between Si and O atoms.

2.2.3.1 L-J Potential

The interaction between the carbon atoms in graphene and the Si, O atoms in the silica substrate was described by van der Waals forces. In addition, the carbon atoms in the graphene also exist a van der Waals force between molecules in long-range. In this study, the general 12/6 potential energy form was selected to describe the van der Waals force between different atoms in the simulation system, in which the substrate and graphene were generally considered as electrostatic balance system. Thus the intermolecular Coulomb's potential contact was ignored. The van der Waals intermolecular force was expressed as:

$$E_{ij}^{LJ} = 4\varepsilon \left[\left(\frac{\sigma_{ij}}{r}\right)^{12} - \left(\frac{\sigma_{ij}}{r}\right)^{6} \right] \quad r < r_c \qquad (2.1)$$

In Eq. (2.1), ε is the energy parameter and σ is the distance parameter.

2.2 Introduction of MD Simulation

In order to save the simulation time, it is generally necessary to set a cutoff distance r_c for the force, and the interaction between the atoms outside the cutoff distance is very weak and negligible. Specific calculation parameters and cutoff radius will be introduced for different models in subsequent simulations.

2.2.3.2 AIREBO Potential

For the interaction between carbon atoms in graphene, the AIREBO was selected. This potential energy function can accurately describe the combination of strong hybrid bonds of sp, sp^2, sp^3 in graphene and the conversion between different hybrid bonds. Meanwhile, it can also accurately describe the brokenness of old chemical bonds and formation of new bonds in the process of graphene joining, doping and nanopore drilling. In addition, it is able to describe the bond energy, bond length, elastic properties, defect formation and surface energy changes during the process of structural transformation. This potential is based on the REBO potential from Brenner, and has been widely used in the study of properties and structure of carbon nanomaterials. The expression of this potential energy is:

$$E = \frac{1}{2} \sum_i \sum_{j \neq i} \left[E_{ij}^{REBO} + E_{ij}^{LJ} + \sum_{k \neq i,j} \sum_{l \neq i,j,k} E_{kijl}^{tor} \right] \quad (2.2)$$

The first term on the right side of (2.2) is the covalent bond (REBO) term, the second term is the intermolecular long-range action term (L-J), and the third term is the torsion item describing the bond rotation. The models in this paper only considered the first two items, for which L-J item was already mentioned in Sect. 2.2.3.1, REBO item can be expressed as:

$$E_{ij}^{REBO} = V_{ij}^R + b_{ij} V_{ij}^A \quad (2.3)$$

In which, V_{ij}^R and V_{ij}^A represent the exclusion and attraction items, respectively:

$$V_{ij}^R = w_{ij}(r_{ij}) \left[1 + \frac{Q_{ij}}{r_{ij}} \right] A_{ij} e^{-\alpha_{ij} r_{ij}} \quad (2.4)$$

$$V_{ij}^A = -w_{ij}(r_{ij}) \sum_{n=1}^{3} B_{ij}^{(n)} e^{-\beta_{ij}^{(n)} r_{ij}} \quad (2.5)$$

Here, Q is the charge, r is the atomic distance, w limits the interaction range, which is expressed as:

Table 2.2 Main function parameters of C-C bonds (AIREBO potential)

Parameters	Value	Parameters	Value
$B1$	12,388.79197798 eV	Q	0.3134602960833 Å
$B2$	17.56740646509 eV	A	10,953.544162170 eV
$B3$	30.71493208065 eV	α	4.7465390606595 Å^{-1}
$\beta1$	4.7204523127 Å^{-1}	D_{min}	1.7 Å
$\beta2$	1.4332132499 Å^{-1}	D_{max}	2.0 Å
$\beta3$	1.3826912506 Å^{-1}		

$$w_{ij}(r_{ij}) = \begin{cases} 1 & r < D_{ij}^{min} \\ 1 + \cos\left(\frac{r - D_{ij}^{min}}{D_{ij}^{max} - D_{ij}^{min}}\right) & D_{ij}^{min} < r < D_{ij}^{max} \\ 0 & D_{ij}^{max} < r \end{cases} \quad (2.6)$$

For the C-C bonds, some of the main parameters are shown in Table 2.2:

2.2.3.3 Tersoff Potential

Tersoff potential function can consider the influence of covalent bond, atomic local environment and bond angle and other factors on the bond order, so it has a good description of the formation and break of covalent bond. Tersoff potential function is widely used in the interaction between carbon and silicon materials, and can also be used to simulate the interaction of carbon-nitrogen, carbon-boron, and silicon-oxygen atoms. In this paper, the Tersoff potential function was used to describe the interaction between Si-O, C-N and C-B atoms. The potential energy function is expressed as:

$$E = \frac{1}{2} \sum_i \sum_{j \neq i} V_{ij} \quad (2.7)$$

$$V_{ij} = f_C(r_{ij})[f_R(r_{ij}) + b_{ij}f_A(r_{ij})] \quad (2.8)$$

In the formula (2.8), r_{ij} represents the distance between atoms, and the interaction termination function f_c can be expressed as:

$$f_C(r) = \begin{cases} 1 & : \quad r < R - D \\ \frac{1}{2} - \frac{1}{2}\sin(\frac{\pi}{2}\frac{r-R}{D}) & : \quad R - D < r < R + D \\ 0 & : \quad r > R + D \end{cases} \quad (2.9)$$

2.2 Introduction of MD Simulation

f_R describes the two-body effect:

$$f_R(r) = A \exp(-\lambda_1 r) \tag{2.10}$$

f_A describes the multi-body action:

$$f_A(r) = -B \exp(-\lambda_2 r) \tag{2.11}$$

b_{ij} is a factor that characterizes the intensity of the multibody effect, which can be expressed as:

$$b_{ij} = (1 + \beta^n \varsigma_{ij}^n)^{-\frac{1}{2n}} \tag{2.12}$$

$$\varsigma_{ij} = \sum_{k \neq i,j} f_C(r_{ik}) g(\theta_{ijk}) \exp[\lambda_3^m (r_{ij} - r_{ik})^m] \tag{2.13}$$

$$g(\theta_{ijk}) = \gamma_{ijk}(1 + \frac{c^2}{d^2} - \frac{c^2}{[d^2 + (\cos\theta - \cos\theta_0)^2]}) \tag{2.14}$$

θ_{ijk} is the bond angle between i-j and j-k, and the remaining parameter values are given in the associated potential energy function file.

2.2.3.4 ZBL Exclusion Potential

The collision force between the incident ions and the graphene was described by the collision energy (ZBL), which uses the transfer function to decay the energy and force into zero within the cut-off range. For the collision between Si, N, B atoms and C atoms in graphene, the ZBL potential function was applied based on the Tersoff potential function, and a Fermi function was adopted to smoothly connect the ZBL exclusion potential and the Tersoff potential energy at the position near the nucleus. The expression is:

$$E = \frac{1}{2} \sum_i \sum_{j \neq i} V_{ij} \tag{2.15}$$

$$V_{ij} = (1 - f_F(r_{ij})) V_{ij}^{ZBL} + f_F(r_{ij}) V_{ij}^{Tersoff} \tag{2.16}$$

where f_F is the Fermi function that connects Tersoff and ZBL potentials, which is expressed as:

$$f_F(r_{ij}) = \frac{1}{1 + e^{-A_F(r_{ij} - r_C)}} \tag{2.17}$$

In Eq. (2.17), r_C is the cutoff radius of ZBL potential. The Tersoff potential section was already described in Sect. 2.2.3.3, and the potential energy of the ZBL exclusion section can be expressed as:

$$V_{ij}^{ZBL} = \frac{1}{4\pi\varepsilon_0} \frac{Z_1 Z_2 e^2}{r_{ij}} \phi(r_{ij}/a) \tag{2.18}$$

$$a = \frac{0.8854 a_0}{Z_1^{0.23} + Z_2^{0.23}} \tag{2.19}$$

$$\phi(x) = 0.1818 e^{-3.2x} + 0.5099 e^{-0.9423x} + 0.2802 e^{-0.4029x} + 0.02817 e^{-0.2016x} \tag{2.20}$$

where Z_1 and Z_2 represent the number of nucleus electrons in the collision, e is the electronic valence, ε_0 is the vacuum dielectric constant, a_0 is the Bohr radius, and the remaining coefficients were given by the original literature.

For the interaction between inert gases like argon and carbon atoms, only the ZBL exclusion potential was used.

2.2.4 Integral Algorithm

MD simulation is to discretize the motion equation in time and space, and use the finite difference equation to solve the motion state of the particles in the time domain. In this process, the time domain is discretized by a finite grid node, and the spacing of adjacent grid points is defined as the time step Δt. The basic process of the solution of MD is to bring the position and velocity of atoms at time t into the Newtonian motion equations to obtain the atomic state at time t + Δt. This process involves a large number of numerical iterations. In order to solve this problem, many different types of integration algorithms were proposed and developed, among which the Verlet algorithm is proved to have good accuracy and stability, and the calculation process is relatively simple. So it's the most widely used algorithm in MD simulation. In this paper, Verlet type algorithm was chosen to do the numerical integration.

The basic idea of the Verlet algorithm is to Taylor expand $\vec{r}_i(t + \Delta t)$ and $\vec{r}_i(t - \Delta t)$ at t:

$$\vec{r}_i(t + \Delta t) = \vec{r}_i(t) + \frac{d\vec{r}_i(t)}{dt}\Delta t + \frac{1}{2}\frac{d^2\vec{r}_i(t)}{dt^2}\Delta t^2 + \frac{1}{3!}\frac{d^3\vec{r}_i(t)}{dt^3}\Delta t^3 + O(\Delta t^4) \tag{2.21}$$

2.2 Introduction of MD Simulation

$$\vec{r}_i(t - \Delta t) = \vec{r}_i(t) - \frac{d\vec{r}_i(t)}{dt}\Delta t + \frac{1}{2}\frac{d^2\vec{r}_i(t)}{dt^2}\Delta t^2 - \frac{1}{3!}\frac{d^3\vec{r}_i(t)}{dt^3}\Delta t^3 + O(\Delta t^4)$$
(2.22)

Adding (2.21) and (2.22) together, and ignoring the higher order terms of Δt^4 and above, we can get:

$$\vec{r}_i(t + \Delta t) = 2\vec{r}_i(t) - \vec{r}_i(t - \Delta t) + \frac{d^2\vec{r}_i(t)}{dt^2}\Delta t^2 \quad (2.23)$$

Subtracting (2.22) from (2.21), and ignoring the high order of Δt^3 and above, we can get:

$$\frac{d\vec{r}_i(t)}{dt} = \frac{\vec{r}_i(t + \Delta t) - \vec{r}_i(t - \Delta t)}{2\Delta t} \quad (2.24)$$

So the basic form of Verlet algorithm is:

$$\begin{cases} \vec{r}_i(t + \Delta t) = 2\vec{r}_i(t) - \vec{r}_i(t - \Delta t) + \vec{a}_i(t)\Delta t^2 \\ \vec{v}_i(t) = \frac{\vec{r}_i(t + \Delta t) - \vec{r}_i(t - \Delta t)}{2\Delta t} \end{cases} \quad (2.25)$$

It can be seen from function (2.25) that due to the difference in the truncation error, the calculation of the velocity in the Verlet algorithm is behind the calculation of the position. In order to obtain the position and speed at the same time, some improved Verlet integral methods were developed. The Velocity-Verlet algorithm can calculate the velocity, position and acceleration at the same time, and the calculation precision is relatively high and the calculation time is moderate. So the Velocity-Verlet algorithm is widely used, and also chosen by this paper. The basic form of the Velocity-Verlet algorithm is:

$$\begin{cases} \vec{r}_i(t + \Delta t) = \vec{r}_i(t) + \vec{v}_i(t)\Delta t + \frac{1}{2}\vec{a}_i(t)\Delta t^2 \\ \vec{v}_i(t + \Delta t) = \vec{v}_i(t) + \frac{1}{2}(\vec{a}_i(t) + \vec{a}_i(t + \Delta t))\Delta t \end{cases} \quad (2.26)$$

Verlet algorithm increases the speed of calculation by sacrificing a part of the calculation accuracy. In order to ensure the accuracy of the calculation of energy, this paper used a small time step. In the subsequent MD simulation, if there is no special illustration, the calculation time step was selected as 0.1 fs.

2.2.5 Simulation Ensemble

In order to more truly reflect the actual physical process, the classic MD simulation is often carried out under certain ensemble conditions. The ensemble is a collection of independent systems that have exactly the same structure and properties and exist in a variety of motion states. The ensembles used in this paper are describes as following:

1. NVE

If the particles of the system are evolved along a constant energy orbit during the course of the movement (energy is fixed), and the particle number N and volume V of the system remain constant, then the system is in the NVE. It can be seen from the definition that NVE is isolated and conservative system.

2. NVT

NVT is the system with constant particle number N, constant temperature T and constant volume V. In order to keep the system temperature constant, a hot bath from outside can be used to coexist with the system, so that it is in thermal equilibrium condition. Also the constant temperature can be achieved by velocity scaling.

3. NPT

In the NPT, the system has a defined temperature T, pressure P and particle number N. To achieve the constant temperature, the adjustment method is the same as NVT ensemble. The adjustment of the pressure is more complex, it is often realized through the system volume scaling.

2.2.6 Averaging of Statistical Results

In this paper, MD was used to explain the phenomena of graphene irradiated by particle beams, and also describe the mechanical properties of graphene after irradiation. Therefore, in addition to the need for describing the dynamic variation of atomic structure in nano-scale, it is also needed to extract some other macro physical quantities, such as temperature field distribution of graphene structure, energy change of the system, the system stress field distribution, and so on. The following describes the extraction method of main physical quantities.

2.2.6.1 Energy

In MD, the total energy of the system contains both kinetic energy and potential energy:

$$E(t) = E_k(t) + U(t) \tag{2.27}$$

where U(t) is solved by the potential energy function described above, and the kinetic energy is calculated from the following equation:

$$E_k(t) = \left(\frac{1}{2}\sum_{i=1}^{N} m_i v_i^2(t)\right) \quad (2.28)$$

2.2.6.2 Temperature

The temperature field in the MD simulation is for a group of atoms, which is converted from the kinetic formula of the system. For a system with N particles in three-dimensional case, the kinetic energy is:

$$E_k = \frac{3N}{2} k_B T \quad (2.29)$$

By converting Eq. (2.29), the expression of the system temperature can be derived as:

$$T = \frac{2}{3Nk_B} E_k = \frac{1}{3Nk_B} \left\langle \sum_i m_i v_i^2 \right\rangle \quad (2.30)$$

2.2.6.3 Stress

In this paper, the analysis of the mechanical properties of graphene involves the extraction of single atom stress and the calculation of the stress and strain of the whole system. The extraction method of single atom stress is as follows:

$$\begin{aligned}
S_{ab} = &-[mv_a v_b + \frac{1}{2}\sum_{n=1}^{N_p}(r_{1a}F_{1b} + r_{2a}F_{2b}) + \frac{1}{2}\sum_{n=1}^{N_b}(r_{1a}F_{1b} + r_{2a}F_{2b}) \\
&+ \frac{1}{3}\sum_{n=1}^{N_a}(r_{1a}F_{1b} + r_{2a}F_{2b} + r_{3a}F_{3b}) \\
&+ \frac{1}{4}\sum_{n=1}^{N_d}(r_{1a}F_{1b} + r_{2a}F_{2b} + r_{3a}F_{3b} + r_{4a}F_{4b}) \\
&+ \frac{1}{4}\sum_{n=1}^{N_i}(r_{1a}F_{1b} + r_{2a}F_{2b} + r_{3a}F_{3b} + r_{4a}F_{4b}) + Kspace(r_{ia}, F_{ib}) + \sum_{n=1}^{N_f} r_{ia}F_{ib}]
\end{aligned} \quad (2.31)$$

In the formula (2.31), assigning x, y, z to a, b can get the corresponding six components stress tensor, respectively. The right side of the equation corresponds to the kinetic energy, pair potential, bonding, body angle, dihedral angle, uncertain item, coulomb term, and binding term.

Note that the stress calculated in Eq. (2.31) is actually the result of the atomic stress multiplied by the volume. The true single-atom stress should be the calculated value divided by the single-atom volume, but it is impossible to calculate the volume of a single carbon atom during the tensile deformation. Therefore, the atomic stress distribution in this paper is actually the result of the stress multiplied by the volume. But it does not affect the distribution of the atomic stress, so this treatment is reasonable.

For the tensile stress of the whole system, the stress of all the atoms is summed and divided by the volume V of the system after the single-atom stress is calculated by using the formula (2.31). That is, the stress value of the system during stretching is:

$$\sigma = \frac{\sum_{i=1}^{N} S_{ab}}{V} \tag{2.32}$$

2.2.7 Introduction of Simulation Software

The MD simulation software used in this paper is LAMMPS [5], which is a parallel MD code developed by Sandia National Laboratory. LAMMPS is widely used to simulate the structure and thermodynamic properties of carbon nanomaterials. Its reliable algorithms and parallel processors ensure that the process of particle beam irradiation of graphene is truly reproduced on a computer.

2.3 Electronic Transport Theory

2.3.1 Introduction

In this paper, the electronic transport properties of graphene before and after particle beam irradiation were investigated by the methods of DFT and NEGF. In this section, the basic theorem of DFT was introduced first, and then the detail form of NEGF was described. After that, the process of solving the electronic transport properties of graphene under DFT and NEGF framework was depicted. Finally, the software for calculating the transport properties was given.

2.3.2 DFT

In the traditional quantum theory, in order to solve the electronic structure of matter, usually the electronic wave function is taken as the basic physical quantity, and then by solving the electronic Schrodinger equation, the basic physical properties are determined. For complex systems, solving the Schrodinger equation for multi-electron system is confronted with large data processing, which is impossible

2.3 Electronic Transport Theory

to be handled accurately at present. With the development of computational materials science, there are a variety of reasonable approximations and simplifications gradually evolved. The basic idea of the modern first-principles method is firstly transferring the multi-body problem to the multi-electron system by the Born-Oppenheimer approximation, and then using the Hartree-Fock approximation and DFT theory to transfer the multi-electron system problem to the single electron system problem, for which the single electron moves in the average potential of the other nuclei and electrons.

2.3.2.1 Born-Oppenheimer and Hartree-Fock Approximation

To solve the electronic energy level of the solid system, we need to solve the Schrodinger equation:

$$H(\vec{r}, \vec{R})\Psi(\vec{r}, \vec{R}) = E\Psi(\vec{r}, \vec{R}) \qquad (2.33)$$

where E and Ψ are the energy eigenvalues and wave functions describing the system, H is the Hamiltonian of the system, and $\vec{r}$ and $\vec{R}$ is the set of coordinates for all the electrons and nuclei in the system, respectively. If the influence of the external field is neglected, the H of the system is expressed as several items: the electron kinetic energy and the coulomb potential energy induced by the interaction between different electrons, the nucleus kinetic energy and the coulomb potential energy induced by the interactions between nucleuses, the interaction between the electron and the nucleus:

$$\begin{aligned} H(\vec{r}, \vec{R}) = & -\sum_i \frac{\hbar^2}{2m}\nabla^2_{r_i} + \frac{1}{2}\sum_{ii'} \frac{e^2}{|\vec{r_i} - \vec{r_{i'}}|} \\ & -\sum_j \frac{\hbar^2}{2M_j}\nabla^2_{R_j} + \frac{1}{2}\sum_{jj'} V_n(\vec{R_j} - \vec{R_{j'}}) - \sum_{ij} V_{e-n}(\vec{r_i} - \vec{R_j}) \end{aligned} \qquad (2.34)$$

It is very difficult to solve the equations of electron and atomic nucleus of the multi-body system. Born-Oppenheimer approximation refers to the separation of electrons and nuclei in the consideration of multibody problems, that is, when studying the electron motion, the nuclei is regarded as static potential disturbance, and when studying the movement of nuclei, neglecting the distribution of electron in space. So that electron and nuclei only need to meet their respective motion equations. Under adiabatic approximation, the Hamiltonian of the electronic part is:

$$H(\vec{r}, \vec{R}) = -\sum_i \frac{\hbar^2}{2m}\nabla^2_{r_i} + \frac{1}{2}\sum_{ii'} \frac{e^2}{|\vec{r_i} - \vec{r_{i'}}|} - \sum_{ij} V_{e-n}(\vec{r_i} - \vec{R_j}) \qquad (2.35)$$

As the weight and speed of movement of electron and atomic nucleus are largely different, reasonable results can be obtained by considering the movement of nucleus and electronic separately. The adiabatic approximation can transform the multi-body solution system into a multi-electron system. But solving of the equation of multi-electron system requires the Coulomb interaction term between electrons, which cannot be accurately determined. Hartree proposed to simplify the interaction between electrons as a single electron in the average potential field of other electrons (single electron approximation). Through this approximation, the motion state of multi-electron system is described with a single electron wave function, and multi-electron system wave function is taken as the product of the single electron wave function. Subsequently, Fock proposed to change the product of the wave function into a Slater determinant in consideration of the spin state. Finally, the single-electron equation of the system is:

$$E_i \Psi_i(\vec{r}) = \left[-\frac{\hbar^2}{2m} \nabla^2 - \sum_j V_{e-n}(\vec{R_j}) \right] \Psi_i(\vec{r}) + \sum_{i' \neq i} \int d\vec{r'} \frac{|\Psi_{i'}(\vec{r'})|^2}{|\vec{r_i} - \vec{r_{i'}}|} \Psi_i(\vec{r})$$
$$+ \sum_{i' \neq i} \int dr' \frac{\Psi_{i'}^* \Psi_i(\vec{r'})}{|\vec{r_i} - \vec{r_{i'}}|} \Psi_{i'}(\vec{r})$$

(2.36)

The single electron wave function is obtained by solving the single electron equation, and then the multi-electron wave function of the system is obtained by the single electron wave function. Finally, the Schrodinger equation of the multi-body system is simplified as a single electron problem by the adiabatic and Hartree-Fock approximation.

2.3.2.2 Thomas-Fermi Model

As early as 1927, Thomas and Fermi proposed the Thomas-Fermi model based on homogeneous electron gas to solve the Schrodinger equation. In this model, the influence of external force on electrons, and the interaction between electrons were ignored. The total energy of the electronic system can be expressed by electron density:

$$E_{TF}[\rho(\vec{r})] = C_{TF} \int \rho(\vec{r})_{5/3} d\vec{r}$$
$$- z \int \frac{\rho(\vec{r})}{\vec{r}} d\vec{r} + \frac{1}{2} \int \int \frac{\rho(\vec{r_1}) - \rho(\vec{r_2})}{|\vec{r_1} - \vec{r_2}|} d\vec{r_1} d\vec{r_2} \quad (2.37)$$

Equation (2.37) shows that the energy functional of the system can be expressed as a function determined only by the electron density. In order to consider the

2.3 Electronic Transport Theory

influence of the exchange correlation between electrons which was ignored by Thomas-Fermi model, Dirac proposed to add a geometric correction term in the formula (2.37), which is the so-called Thomas-Fermi-Dirac model:

$$E_{TF}[\rho(\vec{r})] = C_1 \int d^3\vec{r} \rho^{\frac{5}{3}}(\vec{r}) + \int d^3\vec{r} V_{ext}(\vec{r})\rho(\vec{r})$$
$$+ C_2 \int d^3\vec{r} \rho^{\frac{4}{3}}(\vec{r}) + \frac{1}{2} \int d^3\vec{r} d^3\vec{r}' \frac{\rho(\vec{r})\rho(\vec{r}')}{|\vec{r}-\vec{r}'|} \qquad (2.38)$$

2.3.2.3 Hohenberg-Kohn Theorem

According to the above Thomas-Fermi-Dirac model, Hohenberg and Kohn established the DFT. Strict DFT is based on two important theorems. Among them, Hohenberg-Kohn's first theorem states that the ground state energy of the system is a unique function determined only by the electron density function. It pointed out that the ground state energy of the multi-electron system and the related physical quantities and properties can be uniquely determined by the ground state electron density. The work of solving the Hamiltonian is greatly simplified by introducing the charge density function. Hohenberg-Kohn's second theorem stated that the ground state energy can be obtained by the process of minimizing the system energy based on the ground state electron density. Variation of the system energy based on electron density can determine the ground state energy, so as to obtain the other relevant physical quantities.

Under the two theorems of Hohenberg-Kohn, the energy of the system can be expressed as:

$$E[\rho] = \int d\vec{r} V_{ext}(\vec{r}) d\vec{r} + F[\rho] \qquad (2.39)$$

To solve the ground state energy is to find the variation of $F[\rho]$, which can be represented by kinetic energy, Coulomb term, and the exchange correlation items in the form of expression:

$$F[\rho] = T[\rho] + \frac{1}{2} \int \int d\vec{r} d\vec{r}' \frac{\rho(\vec{r})\rho(\vec{r}')}{|\vec{r}-\vec{r}'|} + E_{xc}[\rho] \qquad (2.40)$$

Therefore, according to the Hohenberg-Kohn theorem, through taking the electron density of the system as the basic variable, the physical properties of the system in ground state can be solved by making the energy functional variation based on the electron density. From Eq. (2.40), it is necessary to determine three problems in order to solve the ground state energy of the system: Determining the electronic density function $\rho(\vec{r})$ of the system, determining the kinetic energy functional $T[\rho]$ of the system, and determining the exchange-correlation functional $E_{xc}[\rho]$.

2.3.2.4 Kohn-Sham Model

Under the framework of Hohenberg-Kohn theory, making the energy functional variance of the system can get:

$$\int d\vec{r}\,\delta\rho(\vec{r}) \left[\frac{\delta T[\rho(\vec{r})]}{\delta\rho(\vec{r})} + V(\vec{r}) + \int d\vec{r}' \frac{\rho(\vec{r}')}{|\vec{r}-\vec{r}'|} + \frac{\delta E_{xc}[\rho(\vec{r})]}{\delta\rho(\vec{r})} \right] = 0 \quad (2.41)$$

Under $\int d\vec{r}\,\delta\rho(\vec{r}) = 0$, we can get:

$$\frac{\delta T[\rho(\vec{r})]}{\delta\rho(\vec{r})} + V(\vec{r}) + \int d\vec{r}' \frac{\rho(\vec{r}')}{|\vec{r}-\vec{r}'|} + \frac{\delta E_{xc}[\rho(\vec{r})]}{\delta\rho(\vec{r})} = \mu \quad (2.42)$$

In the effective potential field, the above equation can be expressed as:

$$V_{eff}(\vec{r}) = V(\vec{r}) + \int d\vec{r}' \frac{\rho(\vec{r}')}{|\vec{r}-\vec{r}'|} + \frac{\delta E_{xc}[\rho(\vec{r})]}{\delta\rho(\vec{r})} \quad (2.43)$$

Due to the kinetic energy function $T[\rho]$ is unknown, in order to solve Eq. (2.41), Kohn-Sham proposed to replace $T[\rho]$ by a kinetic energy functional without considering the interaction term $T_s[\rho]$. The difference between them is considered in exchange-correlation functional. For which, the electronic density function $\rho(\vec{r})$ is solved by a single electron wave function:

$$\rho(\vec{r}) = \sum_{i=1}^{N} |\Psi_i(\vec{r})|^2 \quad (2.44)$$

Thus the Kohn-Sham equation is obtained:

$$\{-\nabla^2 + V_{ks}[\rho(\vec{r})]\}\Psi_i(\vec{r}) = E_i\Psi_i(\vec{r}) \quad (2.45)$$

$$\begin{aligned} V_{ks}[\rho(\vec{r})] &= V(\vec{r}) + V_{coul}[\rho(\vec{r})] + V_{xc}[\rho(\vec{r})] \\ &= V(\vec{r}) + \int d\vec{r}' \frac{\rho(\vec{r}')}{|\vec{r}-\vec{r}'|} + \frac{\delta E_{xc}[\rho(\vec{r})]}{\delta\rho(\vec{r})} \end{aligned} \quad (2.46)$$

For the above equation, the first term is the atomic nucleus attracting potential, the second term is the electron-electron interaction Coulomb term, and the third term is the exchange-correlation part. By solving the single-electron equation, the ground state performance of the system can be obtained.

2.3 Electronic Transport Theory

2.3.2.5 Exchange-Correlation Functional

A large number of unknown parts of the Kohn-Sham equation are assigned to the exchange-correlation part. After determining the electronic density $\rho(\vec{r})$ and kinetic energy $T[\rho]$ of the system, the specific form of the exchange-correlation part must be obtained. There are different methods proposed to deal with the exchange-correlation part, among which LDA and GGA are mostly used. The basic idea of LDA is to use the density function of more uniform electron gas to get the exchange-correlation form of non-uniform electron gas, which is suitable for the system with slow change of electron density. GGA is more applicable for a system with large nonuniformity of the electron density or nonuniformity of the electron density plays an important role to the results. GGA is a functional considering the exchange-correlation energy as the electron charge density and its gradient. It has greatly improved the results of atomic exchange and correlation energy for many systems. However, the GGA method cannot obtain the accurate results for all systems, and the GGA method greatly increases the computational complexity. Therefore, in the actual process it is needed to select the appropriate methods based on system studied and the problems concerned.

2.3.3 Green Function Theory

Usually all the properties of the quantum mechanics system are based on the Hamiltonian of the system, and the eigenfunctions and eigenvalues of the Hamiltonian system are obtained by solving the Schrodinger equation. The solution of Schrodinger equation needs to theoretically construct the wave function. While the solving of the wave function of an actual complex system is impossible, this makes people continually seek new methods and theories to get rid of the solution of wave function to get the system ground state energy, eigenvalue and other physical quantities. Green function method is to describe the system from the particle movement, to avoid solving the wave function to get the relevant physical properties. The physical quantities obtained by solving the Hamiltonian can also be obtained by solving the Green's function. According to the different processing systems, Green's function is divided into equilibrium Green's function and NEGF.

2.3.3.1 NEGF

The equilibrium Green function is defined on the real time axis, and the NEGF is defined on the contour of the complex time plane. The contour-ordered Green function defined in the complex time plane is:

$$G^C(1, 1') \equiv -i\langle T_C \psi(1) \psi^\dagger(1') \rangle \tag{2.47}$$

where $(1) \equiv (r_1, t_1)$ and $(1') \equiv (r_{1'}, t_{1'})$ are abbreviations of the spatial and temporal variables, respectively. The line C starts from $-\infty$ and ends with ∞, passing through the real time axis at time t_0. The upper and lower paths of the real time axis are labeled as C_1 and C_2, respectively. T_C is a complex weave operator defined on the contour. For two-time scale on the contour, defining the following Green function:

$$G^C(1, 1') \equiv \begin{cases} G^t(1, 1') & t_1, t_{1'} \in C_1 \\ G^>(1, 1') & t_1 \in C_1, t_{1'} \in C_2 \\ G^<(1, 1') & t_1 \in C_2, t_{1'} \in C_1 \\ G^{\tilde{t}}(1, 1') & t_1, t_{1'} \in C_2 \end{cases} \tag{2.48}$$

Among them, the time series Green function $G^t(1, 1')$, greater Green function $G^>(1, 1')$, less Green function $G^<(1, 1')$, anti-temporal Green function $G^{\tilde{t}}(1, 1')$, retarding Green function $G^R(1, 1')$, advancing Green function $G^A(1, 1')$ are written as:

$$G^t(1, 1') \equiv -i\theta(t_1 - t_{1'})\langle \psi(1)\psi^\dagger(1') \rangle + i\theta(t_{1'} - t_1)\langle \psi^\dagger(1')\psi(1) \rangle \tag{2.49}$$

$$G^>(1, 1') \equiv -i\langle \psi(1)\psi^\dagger(1') \rangle \tag{2.50}$$

$$G^<(1, 1') \equiv -i\langle \psi^\dagger(1')\psi(1) \rangle \tag{2.51}$$

$$G^{\tilde{t}}(1, 1') \equiv -i\theta(t_{1'} - t_1)\langle \psi(1)\psi^\dagger(1') \rangle + i\theta(t_1 - t_{1'})\langle \psi^\dagger(1')\psi(1) \rangle \tag{2.52}$$

$$G^R(1, 1') = \theta(t_1 - t_{1'})[G^>(1, 1') - G^<(1, 1')] \tag{2.53}$$

$$G^A(1, 1') = \theta(t_{1'} - t_1)[G^<(1, 1') - G^>(1, 1')] \tag{2.54}$$

The six Green functions are related to each other. In order to solve the non-equilibrium state, only four Green functions need to be known. Among them, the retarding and the advancing Green function describe the energy spectrum, the state density, etc. The greater and less Green functions describe the particle number information in the system.

The Dyson equation can consider the self-energy, the Green's function and the undisturbed Green's function all together. The Dyson equation is:

2.3 Electronic Transport Theory

$$G^C(1,1') = G_0^C(1,1') + \int dx_2 \int_C d\tau_2 G_0^C(1,2) U(2) G^C(2,1')$$
$$+ \int dx_2 \int dx_3 \int_C d\tau_2 \int_C d\tau_3 G_0^C(1,2) \sum{}^C(2,3) G^C(3,1') \quad (2.55)$$

In the Dyson equation, $G_0^C(1,1')$ is the Green function of a system without any interactions. The external potential energy U includes the non-equilibrium part. Self-energy $\sum^C[G]$ has the interaction part. Therefore, it describes the method of using the non-interaction Green's function to represent the system with interaction.

In solving the non-equilibrium system, Keldysh also proposed the corresponding motion equations:

$$G^R = G_0^R + G_0^R \sum{}^R G^R \quad (2.56)$$

$$G^A = G_0^A + G_0^A \sum{}^A G^A \quad (2.57)$$

$$G^< = (1 + G^R \sum{}^R) + G_0^<(1 + G^A \sum{}^A) + G^R \sum{}^< G^A \quad (2.58)$$

$$G^> = (1 + G^R \sum{}^R) + G_0^>(1 + G^A \sum{}^A) + G^R \sum{}^> G^A \quad (2.59)$$

$$G^t = (1 + G^R \sum{}^R) + G_0^t(1 + G^A \sum{}^A) + G^R \sum{}^t G^A \quad (2.60)$$

$$G^{\tilde{t}} = (1 + G^R \sum{}^R) + G_0^{\tilde{t}}(1 + G^A \sum{}^A) + G^R \sum{}^{\tilde{t}} G^A \quad (2.61)$$

2.3.3.2 NEGF Theory in Electronic Transport

By using the NEGF method, the electron density matrix of the system can be obtained without solving the wave function, which provides an effective method for studying the electronic transport properties of the non-equilibrium open system.

In a system described by Hamiltonian, the Green function is defined as:

$$G(z) = (z\hat{S} - \hat{H})^{-1} \quad (2.62)$$

where z is defined on the complex plane and $\hat{S}$ is the overlapping matrix of the base set. In the study of electronic transport problems, the solution of such systems corresponds to infinite $\hat{H}$ and $\hat{S}$, and the corresponding matrix $(z\hat{S} - \hat{H})$ is also bound to be infinite, so we are facing a problem of solving the inversion of infinite matrix. But in fact, it is possible to obtain the properties of the whole system only by solving the Green function of the intermediate scattering part through shield approximation.

The matrix $(z\hat{S} - \hat{H})$ can be written according to the left and right electrodes and the scattering section:

$$z\hat{S} - \hat{H} = \begin{bmatrix} z\hat{S}_L - \hat{H}_L & \tau_L & 0 \\ \tau'_L & z\hat{S}_d - \hat{H}_d & \tau'_R \\ 0 & \tau_R & z\hat{S}_R - \hat{H}_R \end{bmatrix} \quad (2.63)$$

where L and R denote left and right electrodes, d denotes an intermediate scattering region. $\tau_i = z\hat{S}_{id} - \hat{H}_{id}$ and $\tau'_i = z\hat{S}_{di} - \hat{H}_{di}$ represent the interaction between an electrode and an intermediate scattering region, respectively. While the interaction between the two electrodes can be simplified to zero. For the solution of the Green function G, it can be expressed in block form:

$$\begin{bmatrix} z\hat{S}_L - \hat{H}_L & \tau_L & 0 \\ \tau'_L & z\hat{S}_d - \hat{H}_d & \tau'_R \\ 0 & \tau_R & z\hat{S}_R - \hat{H}_R \end{bmatrix} \begin{bmatrix} G_L & G_{Ld} & G_{LR} \\ G_{dL} & G_d & G_{dR} \\ G_{RL} & G_{Rd} & G_R \end{bmatrix} = I \quad (2.64)$$

From which, G_d can be obtained as:

$$G_d = (z\hat{S}_d - \hat{H}_d - \sum_L(z) - \sum_R(z))^{-1} \quad (2.65)$$

In which, $\sum_i(z)$ is the self-energy of the system:

$$\sum_i(z) = \tau'_i g_i(z) \tau_i \quad (2.66)$$

The electrode Green function $g_i(z)$ is expressed as:

$$g_i(z) = (z\hat{S} - \hat{H}_i)^{-1} \quad (2.67)$$

The density of States in the intermediate extended region can be obtained directly from the spectral function:

$$DOS(E) = \frac{1}{2\pi} Tr[A(E)\hat{S}] = \sum_n \delta(E - E_n) \quad (2.68)$$

Under the equilibrium state, the density matrix can be obtained by integrating the spectral function below the Fermi level. In the case of external bias, the electronic state and density matrices are divided into an equilibrium part and a non-equilibrium part, where the electronic state below the lowest Fermi level μ_R constitutes the density matrix of the equilibrium part, which is expressed as:

2.3 Electronic Transport Theory

$$\rho^{eq} = \frac{1}{2\pi} \int_{E_{min}}^{\mu_R} A(E)dE = \frac{i}{2\pi} \int_{E_{min}}^{\mu_R} (G^R(E) - G^A(E))dE \qquad (2.69)$$

The electronic states between μ_R and μ_L forms a non-equilibrium density matrix. The density matrix of the non-equilibrium part can be obtained by integrating the left spectrum function:

$$\rho^{ne} = \frac{1}{2\pi} \int_{\mu_R}^{\mu_L} A(E)dE = \frac{i}{2\pi} \int_{\mu_R}^{\mu_L} (G^R[\Gamma_L f(E - \mu_L)]G^A dE \qquad (2.70)$$

$\Gamma_L = i(\sum_L - \sum_L^+)$ is the imaginary part of the self-energy of left electrode.

Therefore, the density matrix of the system can be firstly obtained by the NEGF method, and then the other electronic structure properties of the system can be derived. Finally, the electronic transport current of the system can be obtained by Landauer-Büttiker formula:

$$I = \frac{2e}{h} \int [f(E - \mu_L) - f(E - \mu_R)] T(E, V) dE \qquad (2.71)$$

$T(E, V) = Tr\left[\Gamma_L G_d^{R*} \Gamma_R G_d^R\right]$ is the transmission function.

2.3.4 Solution Process of the Electronic Transport Properties of Graphene

In order to solve the electronic transport properties of graphene before and after particle beam irradiation, this thesis adopted the method combining DFT and NEGF theory. The specific solving process is as follows:

1. First, the non-equilibrium system to be solved is divided into three parts, in which the left and right parts are semi-infinite electrode region, and the middle part is the scattering area, as shown in Fig. 2.4. This double electrode model can be used to simulate the contact between the electrode region and the scattering region and the carriers transport process in the intermediate scattering region in actual measurement.
2. Solving the transmission of the system by means of the Green's function using the self-consistent process as shown in Fig. 2.5, and then using the Landauer-Büttiker (Eq. 2.71) to solve the current value between the left and right electrodes.

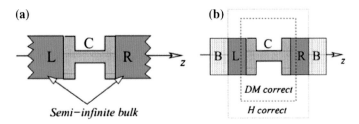

Fig. 2.4 Two-electrode model for calculation of the transport properties of graphene. Reprinted with permission from Ref. [6], Copyright 1998 by American Physical Society

2.3.5 Introduction to Simulation Software

This paper calculated the transport properties of graphene using Transiesta software [6], which combines DFT theory with NEGF theory. It is an open source code developed on the basis of SIESTA software. The non-local conservation pseudopotential was used to describe nuclear electrons and the valence electrons were described by atomic orbital linear combination in finite regions. Combining NEGF method, the electronic transport properties of semi-infinite long electrode molecular devices can be self-consistently calculated, especially under different bias conditions. At present, the commercial software ATK (Atomistix ToolKit) [7] developed based on it has already been widely used in the calculation of the electronic transport properties of CNTs, graphene and other carbon materials. The Transiesta software is very suitable for studying the electronic transport properties of systems with graphene doping, joining and nanoporous structure.

2.4 Chapter Summary

This chapter introduced the experimental and simulation methods used in this research, including the preparation and transfer of graphene materials and the methods of graphene processing and detection. It also included the basic theory of MD simulation and electronic transport calculation. The experiment and the simulation methods used in the subsequent chapters were based on the contents of this chapter, so similar contents in the subsequent sections will not be repeated.

2.4 Chapter Summary

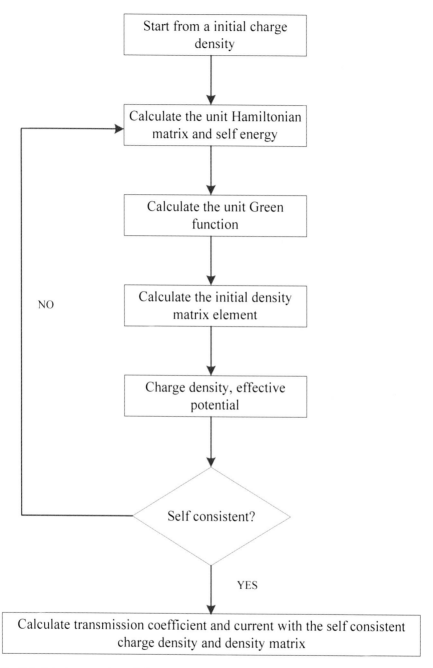

Fig. 2.5 Solution procedure of electron transport properties (DFT + NEGF method)

References

1. Ferrari AC, Meyer JC, Scardaci V et al (2006) Raman spectrum of graphene and graphene layers. Phys Rev Lett 97:187401
2. Stuart SJ, Tutein AB, Harrison JA (2000) A reactive potential for hydrocarbons with intermolecular interactions. J Chem Phys 112:6472–6486
3. Ziegler JF, Biersack JP, Littmark U (1985) The stopping and range of ions in matter. Pergamon, New York
4. Tersoff J (1988) New empirical approach for the structure and energy of covalent systems. Phys Rev B 37:6991
5. Plimpton S (1995) Fast parallel algorithms for short-range molecular dynamcis. J Comput Phys 117:1–19
6. Brandbyge M, Mozos JL, Ordejon P et al (2002) Density-functional method for nonequilibrium electron transport. Phys Rev B 65:165401
7. www.atomistix.com

Chapter 3
General Mechanisms During the Interaction Between Particle Beam and Graphene

3.1 Introduction

Particle beam processing technology is the forefront of the current manufacturing technology, which combines today's high-tech and cutting-edge manufacturing technology. It has many advantages in the application of graphene structure processing. Such as electron beam can achieve a high energy density, and has a good controlled deflection flexibility, thus it can realize graphene processing in different orientations, like nano-hole processing, joining, etc. Particle beams can also be focused to be very fine beams, like micro-scale (laser) or even nano-scale (electron beam) focal spot, which could be able to do the fine processing of graphene materials, such as the selected region doping, nano-hole processing. In order to analyze the process of graphene nanomanufacturing by particle beams irradiation, firstly, the interaction mechanisms between particle beam and graphene need to be figure out, which include the energy conversion between graphene and particle beam under different particle beams irradiation, the changes of graphene structure, graphene damage threshold, and the influence factors of the change of graphene structure. In this chapter, the phenomena and mechanisms of the variations of graphene structure under the action of laser, ion beam and electron beam irradiation were expounded by the combination of theoretical and experimental methods, which lays foundation for the research of the graphene processing by particle beam in later chapters.

3.2 Interaction Between Laser Beam and Graphene

3.2.1 Damage Threshold of Graphene Irradiated by Single Pulse Laser

When the laser irradiates on the surface of graphene, laser photon will generate energy resonance with the electrons and phonons in graphene lattice in a very short period of time. Then, photons will transfer the energy to the carbon atoms in graphene, and the kinetic energy of carbon atoms in graphene will increase due to the energy adsorbed, which could result in the irradiated carbon atoms breaking away from the surrounding atoms. At the same time, the energy can transfer and diffuse in graphene lattice to enlarge the influenced region. In order to study the change of graphene structure under ultrafast laser, in this section, firstly, MD simulation was applied to study the failure threshold of monolayer graphene under single pulse ultrafast laser. The model is shown in Fig. 3.1. For the simulations, LAMMPS was chosen as running program. The interaction between the carbon atoms was described by AIREBO, with a cutoff distance as 2.0 Å. The L-J partial cutoff distance was set as 10.2 Å. The single layer graphene has a size with 10×10 nm, and periodic boundary condition was adopted in the model. Due to the focused spot of ultra-fast laser (micron scale) in experiment was much larger than the established model (nanoscale), the laser was simply taken as heat source directly applied on the graphene to facilitate the study. There were also some researchers [1, 2] using a similar method to investigate the interaction between ultrafast laser and material. The change of the graphene structure under laser with different pulse widths was observed by changing the energy density (the laser absorption rate is 2.3%) and the action time of laser.

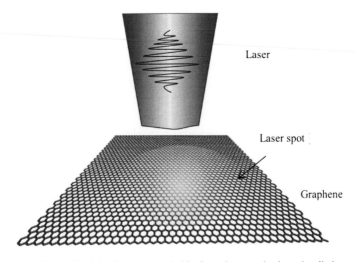

Fig. 3.1 Research model of the damage threshold of graphene under laser irradiation

3.2 Interaction Between Laser Beam and Graphene

Figure 3.2 shows the comparison between calculated ultrafast laser failure threshold of graphene (the threshold was defined as laser energy density in single pulse when graphene is damaged) and the experimental results in Ref. [3]. Under the action of single pulse ultrafast laser, the simulation results are similar to the experimental results: the failure threshold of graphene decreases with the increase of ultrafast laser pulse width. While the simulation results are slightly lower than the experimental results. This is due to the different definitions of failure threshold for simulation and experiment. In the simulation, the failure threshold was defined as the energy density when obvious structural defects (carbon atom sputtering) in the graphene appeared, while in experiment the failure threshold was defined as the energy density when the graphene structure was destructed with a micropore. The energy required to form the pore is larger than that of the structural defect, so that it can be considered that the simulation results and experimental results have a good match. The results also confirmed the correctness of the model used in the subsequent study of laser irradiation to join graphene. Then, it was calculated that when the laser pulse width is 5 ps (the laser pulse width used for the subsequent graphene joining calculation), the failure threshold of graphene is 1.739×10^{10} W cm^{-2}. It was also noted in Ref. [3] that the safe working pulse energy of femtosecond laser should be less than 10^{10} W cm^{-2}. In this work, the laser pulse energy used in the subsequent graphene joining studies is 4.29×10^{8} W cm^{-2}, under which the graphene structure will not be broken.

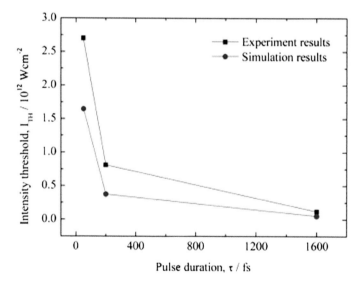

Fig. 3.2 Damage threshold of monolayer graphene under single pulse laser irradiation. Reprinted with the permission from Ref. [4]. Copyright 2013 The Japan Institute of Metals and Materials

3.2.2 The Change of the Morphology of Graphene Under Ultrafast Laser

In general, under the action of laser energy, the lattice will be enlarged, which could macroscopically reflect the thermal expansion of the material. The graphene material is confirmed to have a negative thermal expansion coefficient in a certain temperature range [5]. Thus, graphene may exhibit a reduced volume under the irradiation of laser with energy below damage threshold. According to the single pulse laser damage threshold obtained previously, the change of the morphology and structure of graphene under laser irradiation with energy below damage threshold were simulated. The model size was 5×10 nm (10 nm along the armchair edge), and the laser power was 4.29×10^8 W cm^{-2}, with a laser action time as 5 ps. The other parameters were consistent with Fig. 3.1. Figure 3.3 shows the changes in the overall morphology of graphene under laser irradiation. After the laser energy, the graphene structure will present obvious up and down movement. It was found that the fluctuation of the graphene structure can reach 5.63 Å at the zigzag boundary, and the undulation of the graphene structure can reach 9.15 Å at the armchair boundary. This paper also simulated the structural fluctuation of graphene sheet with the same size at room temperature (300 K). At room temperature, the undulation of graphene structure at the zigzag boundary was about

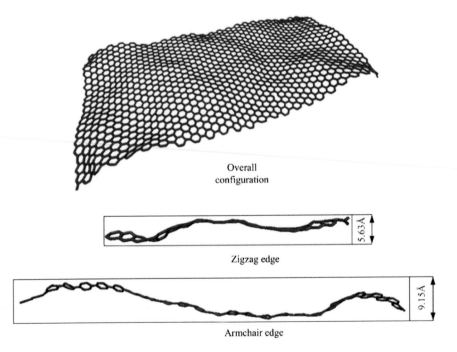

Fig. 3.3 Changes in the morphology of graphene after ultrafast laser irradiation. Reprinted with the permission from Ref. [4]. Copyright 2013 The Japan Institute of Metals and Materials

3.64 Å, and the fluctuation at the armchair-type boundary was about 3.95 Å. Therefore, the thermal effect of laser greatly increases the fluctuation degree of graphene. The lattice of a material will expand under external heat, so will graphene material. However, the thermal action also causes the graphene film to fluctuate perpendicularly to the plane of the graphene. The thermal expansion of the graphene lattice causes an increase of the length in the plane, while the fluctuation of graphene in the perpendicular direction can cause a decrease of the length in the plane. In the case where the temperature is not very high, the fluctuation effect of graphene in the perpendicular direction is greater than the in-plane expansion effect, which could lead to the in-plane negative thermal expansion coefficient reported in Ref. [5].

In order to understand the deformation amplitude of graphene under ultrafast laser irradiation, the movement of carbon atoms at the edge of graphene was further observed. The results were shown in Fig. 3.4. It can be seen that the carbon atoms in the graphene move back and forth in direction parallel to and perpendicular to the plane due to the thermal effect of the laser. In addition, as the temperature increases, the movement of carbon atoms in the plane and perpendicular to the plane will become more intense, even reach 1–2 nm amplitude at higher temperature. While the graphene structure at room temperature has a fluctuation within 0.5 nm under the steady state. The increase in expansion-shrinkage and undulation of the graphene along the plane and in the direction perpendicular to the plane can provide a driving force for the joining of graphene samples, which may overcome the original spacing between the two graphene and lead to the formation of chemical bonds. The joining of graphene samples under ultrafast laser irradiation will be described in Chap. 5.

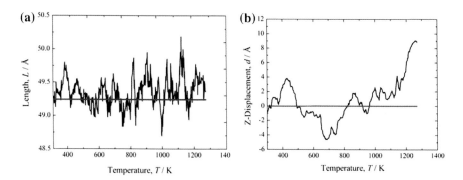

Fig. 3.4 **a** The overall length and **b** Z direction (perpendicular to the graphene plane direction) displacement of the carbon atoms at the edge with temperature changes (red line is set as the baseline) for armchair-type graphene. Reprinted with the permission from Ref. [4]. Copyright 2013 The Japan Institute of Metals and Materials

3.2.3 Experimental Processing of Graphene Structure Under Ultrafast Laser Irradiation

When the energy of ultrafast laser is above the failure threshold, the irradiation of the laser will cause the lattice vibration of the graphene to aggravate, which leads to the destruction of the graphene structure. The laser energy will also transfer and diffuse in graphene lattice, which could reduce the accuracy of laser processing. In this section, the experimental method was used to study the processing effect of laser on graphene with energy above the damage threshold. The laser used in the study was from a pumped mode-locked titanium-doped sapphire laser amplifier with a laser center wavelength of 800 nm, and a pulse width of 50 fs. The maximum pulse frequency was 1000 Hz. The graphene used in the experiment was multilayer (3–5 layers) graphene prepared by CVD. The graphene was transferred to the silicon substrate by chemical etching. The substrate has a 300 nm oxide layer and the focal spot of laser was about 10 μm. By controlling the moving speed of the laser spot in the horizontal and vertical directions, different array patterns can be fabricated on the graphene. In this paper, the laser with energy above the threshold was used to process the graphene microribbons and the micropores array. Figure 3.5 is the experimental setup of graphene processing by ultrafast laser irradiation.

Figure 3.6 shows the optical microscopy results of graphene processed by femtosecond laser. The left side shows the process of processing, using the method of gradually reducing the laser energy to obtain the final failure threshold (defined as the laser energy per unit time when the graphene is removed, note that the definition of damage threshold is different from the simulation). Finally, the single-pulse failure threshold of the multi-layer graphene under the femtosecond

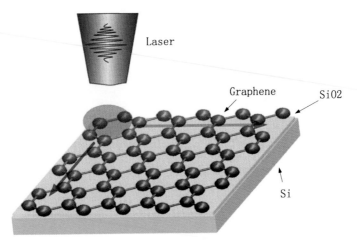

Fig. 3.5 Illustration of the experimental setup of graphene processing by ultra-fast laser irradiation

3.2 Interaction Between Laser Beam and Graphene

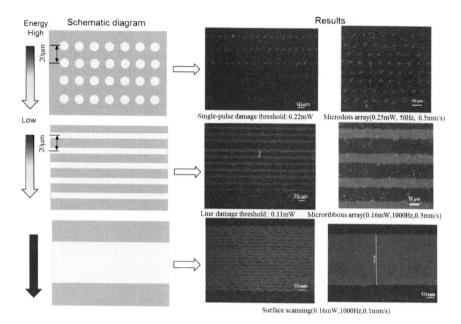

Fig. 3.6 Variation of graphene morphology under laser irradiation with different operating parameters

laser was obtained as 0.22 mW, and the line failure threshold of the graphene was 0.11 mW under the femtosecond laser with a pulse speed as 0.5 mm/s and pulse frequency as 1000 Hz. And then microporous array and microribbon array on graphene were fabricated with laser energy above the damage threshold. We can see that the machining accuracy can be controlled within 10 μm. If the spacing between the microribbons on graphene is reduced, it was found that the laser has an overall blow-off effect on the graphene. By controlling the energy density, the laser may also peel off the graphene layers, resulting in a thinning effect, which has been confirmed in the study of the interaction between picosecond laser and graphene [6].

3.3 Interaction Between Ion Beam and Graphene

3.3.1 The Phenomenon of Graphene Irradiated by Different Energy Ion Beam

Compared with the laser photon, the ion beam particles are stronger in corpuscular property and weaker in volatility. When the ion beam irradiates on graphene, there will more obvious collision effect between atoms, while the thermal diffusion and

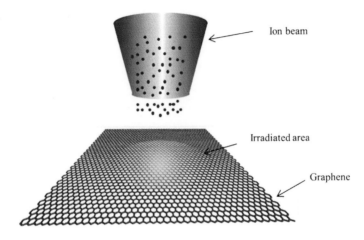

Fig. 3.7 Study model of graphene irradiated by ion beam

heat transfer effect of the collision process will be obviously reduced. In this section, we studied the change of graphene structure under ion beam irradiation. Firstly, the MD method was used to simulate the interaction between ion beam and graphene. The model is shown in Fig. 3.7, in which the incident ions are simplified as non-charged atoms, similar as the method in Ref. [7, 8] dealing with the collision between the ion beam and graphene. The ion beam used in the study includes Ar ion, C ion and Ga ion beams. The interaction potential between carbon atoms was AIREBO, and the interaction potential energy between the incident ions and the carbon atoms in graphene was ZBL. The energy of the incident ions ranges from 20 eV to 1 keV, and the graphene size was 10 × 10 nm. The carbon atoms in the boundary were fixed, and a Berendsen hot bath was applied for the atoms near boundary to absorb the stress waves generated by the shock.

The simulation process is as follows:

(1) The model was relaxed at a temperature of 300 K for a sufficient period of time.
(2) The ions were incident from a region 60 Å above the graphene plane, and an ion was randomly emitted every 500 MD time steps. The ions move in a direction perpendicular to the plane of the graphene and the velocity of the ions can be determined by the applied ion beam kinetic energy. By changing the kinetic energy of the ion beam, the change of graphene structure under the action of different energy ion beam irradiation can be researched. The ion dose was expressed by the number of incident ions per unit area. The NPT ensemble was adopted for the irradiation process.
(3) After the irradiation, the model temperature was increased to 2000 K, and kept the temperature for 10 ps, so that the graphene structure can be relaxed at high temperature for sufficient time.
(4) Cooling the system to room temperature, and maintaining at room temperature to get a steady state. Then the simulation is over.

3.3 Interaction Between Ion Beam and Graphene

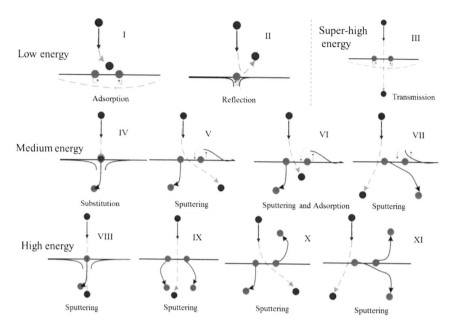

Fig. 3.8 The phenomenon of graphene irradiated by different energy ion beam (red ball represents the incident ion, gray ball represents the carbon atom)

Under the action of ion beam, charged ions will collide with carbon atoms in graphene. According to the different collision energy, there are four phenomena in graphene: ion beam reflection, ion beam adsorption on graphene surface, ion beam embedded within graphene, ion beam transmission. In this paper, the trajectories of the incident ions in graphene were observed in real time and are classified in Fig. 3.8. When the energy is low (usually less than 20 eV), the ions will adsorb on the surface of the graphene, but if the ion beam (Ar^+, Ga^+) has weak interaction with carbon atoms in graphene, the low energy incident ions may be rebounded by graphene. With the increase of ion beam energy (tens eV to several hundred eV), ions will knock the carbon atoms out of graphene. The incident ions may remain in the original location of the sputtered carbon atoms (substitution doping) or pass together with the sputtered carbon atoms, or may remain in the graphene surface to produce adsorption doping, under which the adsorbed ions will migrate to vacancy defects in the subsequent relaxation process. When the carbon atoms in the graphene are sputtered, the carbon atoms near the sputtering atoms are also disturbed by the sputtered atoms and vibrate near its original position. If the energy of the incident ion beam is relatively high, the incident ions or primary sputtered carbon atoms will still have energy to knock the neighbored atoms out of graphene lattice, causing cascade sputtering. When the ion beam energy is very high (thousands eV), the incident ions will pass through the graphene lattice in a second, without producing any sputtering effect to the carbon atoms. Of course, due to the random irradiation of ion beam, the atoms in graphene may receive different kinetic energy

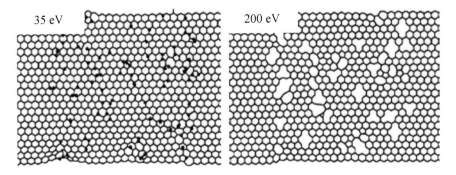

Fig. 3.9 Defects generated in graphene under carbon ion beam irradiation. Reprinted from Ref. [9], Copyright 2014, with permission from Elsevier

due to the difference in energy transfer, which means all the phenomena of graphene induced by different energy ion beam irradiation may occur at one single energy irradiation. The above description stated the main phenomena under the certain energy ion beam irradiation.

The sputtering and adsorption of the incident ion beam will cause defects in the graphene. Figure 3.9 shows the defects of graphene incident by carbon atoms, including single vacancies, double vacancies, polygons, substitution atoms and adsorbed atoms. These defects break the sp^2 hybrid structure in the graphene plane and partially convert it into sp or sp^3 form to introduce unsaturated atoms in the graphene plane. These unsaturated atoms tend to be saturated, which can be used as a driving force for graphene processing, such as graphene joining.

The graphene structure can be processed with different energy and different species of ion beams, depending on the different phenomena produced by ion beam irradiation of graphene. For example, the use of low-energy ion beam can be used to study the implantation doping of graphene, and the use of a medium energy ion beam to study the joining of graphene, or the use of higher energy ion beam to investigate the fabrication of graphene nanopore structure.

3.3.2 Effect of Substrate on Ion Beam Irradiation of Graphene

The actual graphene is generally on a substrate with a certain structure, and the substrate will have an effect on the destruction of graphene structure when irradiated by ion beam. In this section, the influence of the substrate on the irradiation of graphene by ion beam was studied by MD simulation. A SiO_2 substrate, which is common in experiment, was added to the model in Fig. 3.7. Due to limitation of the simulations, the presence of Si under the SiO_2 substrate was neglected, which is reasonable because the Si substrate does not affect the damage of graphene structure

3.3 Interaction Between Ion Beam and Graphene

under relatively low energy ion beam irradiation. In order to match the experimental study, Ga^+ was selected as the incident ions, with an incident energy ranging from 0.1 to 1000 keV. In the MD model, the thickness of SiO_2 was 3.5 nm (too large model is heavily time consuming, and this thickness can cover the focus of this research). The initial spacing between graphene and SiO_2 was 2.9 Å, and the (0001) oxygen atom terminal in the substrate was contacted with graphene. The interaction between the carbon atoms in graphene was described by AIREBO, and the interaction between the silicon atom and oxygen atom was described by the Tersoff/ZBL potential function. ZBL potential function was adopted to represent the collision process between the incident Ga^+ and the C, Si and O atoms. L-J potential was applied to simulate the interaction between Si, O and C atoms, and the parameters of L-J potential were derived from the literature [10].

Figure 3.10 shows the simulation results for graphene incident by 30 keV Ga^+ ion beam. The gallium ion will firstly penetrate the graphene (Fig. 3.10c) and cause defects in the graphene. Since the defects were caused by the direct collision of the gallium ions with the carbon atoms in graphene, they are defined as direct defects, which are mainly consisted of vacancy defects (Fig. 3.10d). At this time, gallium ions still have a high energy, which could further cause the silicon oxygen atoms in substrate to be sputtered with the gallium ion movement (Fig. 3.10e). The sputtered substrate atoms will move in a direction against the incident ions, so they can collide with the graphene on the top of substrate (Fig. 3.10f), which would bring defects into graphene structure once again. At this time, the defects were caused by the sputtering of the atoms in substrate, and they were defined as indirect defects. The indirect defects are also mainly composed of vacancy defects (Fig. 3.10h). Thus, for substrate-supported graphene irradiated by ion beams with large energy, the destruction of graphene was caused by the direct defects resulted from the incident ions and the indirect defects resulted from the atomic sputtering of the substrate. Because the sputtering direction of the substrate atoms under the collision of the incident ions is random, in the case of large-angle sputtering, there will be collision and detachment of the carbon atoms at the location far away from the incident trajectory, which could reduce the processing precision of graphene by FIB irradiation. This also explains the phenomenon that though the focusing radius of the ion beam in the experiment can reach a number below 5 nm, the accuracy of nanopore processing in graphene is much lower than the focusing range.

Then, the possibilities of direct collision defects and indirect collision defects under different energy ion beam incident were studied in order to grasp the mechanism of nanostructure processing in graphene under different energy ion beam irradiation. In this paper, 100 times simulations were conducted for the ion beam incident under each energy, and the number of direct collision defects and indirect collision defects was extracted to calculate the proportion of direct defects and indirect defects occurred under different incident energy, the results are shown in Fig. 3.11. It is seen that the proportion of the collision defect caused by the substrate atoms sputtering (indirect defect) is always greater than the proportion of the direct collision defect, and the difference between them is maximum when the energy ranges from 2 to 10 keV. When the energy of the ion beam is very high, the

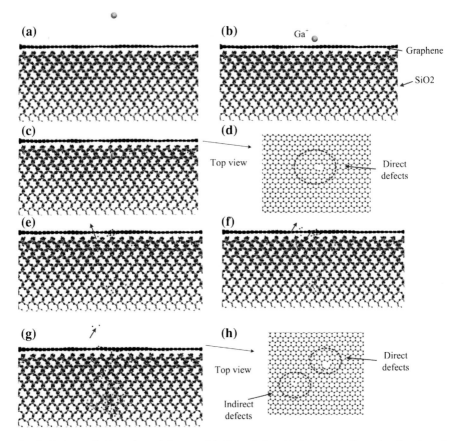

Fig. 3.10 Dynamic formation of direct and indirect defects (red arrows indicate the direction of atomic motion). **a–d** The formation of direct defects; **e–h** The formation of indirect defects. Reprinted with permission from Ref. [11]. Copyright 2015, AIP Publishing LLC

collision cross section of the incident ion beam and graphene is greatly reduced, so that the incident ions penetrate the graphene film without causing any destruction of the graphene structure. Figure 3.8 also shows that when the ion beam energy is very high, there is basically no collision between incident ion beam and graphene, which is consistent with the results here.

In addition, the Monte Carlo simulation method was also used to study the phenomenon of graphene impacted by 30 keV gallium ions. The Monte Carlo method can simulate larger system and incident ion with higher dose, so the characteristics of graphene irradiated by ion beam can be described from a more macroscopic point of view. In the Monte Carlo simulation, in order to reproduce the experimental process, both single-layer and multi-layer graphene were used and 300 nm thickness of the silicon oxide layer was taken as substrate (incident ions cannot reach the thickness of the silicon layer, so the silicon layer was ignored).

3.3 Interaction Between Ion Beam and Graphene

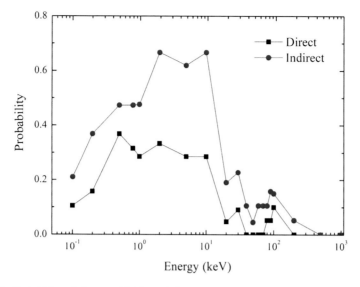

Fig. 3.11 Probability of direct and indirect defects under different energy ion beams irradiation. Reprinted with permission from Ref. [11]. Copyright 2015, AIP Publishing LLC

Figures 3.12 and 3.13 depict the sputtering of carbon atoms in graphene and Si, O atoms in substrate under Ga^+ ion beam irradiation with different energy and dose. As can be seen from Fig. 3.12a, the number of sputtered carbon atoms starts to increase as the energy of the incident ions increases, which is because higher energy ions can transfer more energy into the graphene structure. However, when the ion beam energy increases to a certain value, the collision cross section of the incident ions will be reduced, so that some incident ions will penetrate through graphene. At this time, the number of sputtered carbon atoms will decrease with the increase of the incident ion energy, which is consistent with MD simulation results. In addition,

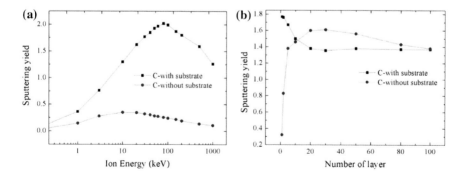

Fig. 3.12 Effect of **a** energy and **b** graphene layer on carbon atom sputtering in graphene during ion beam irradiation. Reprinted with permission from Ref. [11]. Copyright 2015, AIP Publishing LLC

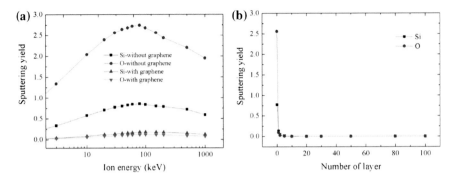

Fig. 3.13 Effect of **a** energy and **b** graphene layer on the sputtering of Si and O atoms in the substrate during the ion beam irradiation. Reprinted with permission from Ref. [11]. Copyright 2015, AIP Publishing LLC

the presence of the substrate greatly increases the number of carbon atoms sputtered. The influence of the number of graphene layers on the number of sputtered carbon atoms in Fig. 3.12b also reflects the effect of the substrate (the presence of the substrate has a greater effect on graphene with few layers and this effect gradually decreases to zero with the increase of the number of graphene layers). Figure 3.13a shows that the variation trend of the number of sputtered silicon and oxygen atoms is similar to that of carbon atoms, except that number of sputtered atoms is significantly smaller than that of carbon atoms, which is due to the fact that the sputtered surface substrate atoms are mostly rebounded by the graphene. If the graphene is removed, it can be found that the number of sputtered substrate atoms greatly increases, indicating that graphene has a significant impact protection effect. The results in Fig. 3.13b show that monolayer graphene can block the sputtering of most of the substrate atoms. The results of this study suggest the potential of monolayer graphene as an impact-resisting material, which is being under investigated.

3.4 Interaction Between Electron Beam and Graphene

In the experiment, the interaction between electron beam and graphene is common. For example, when the graphene sample is characterized by SEM and TEM, on one side, the graphene sample generates an induction signal under the bombardment of the electron beam, which corresponds to the structural morphology and atomic composition of graphene. On the other hand, graphene will also produce structural changes, or even damage under the bombardment. So we need to have a good understand about the interaction between graphene and electron beam to explain the experiment phenomena well. Meanwhile, it is worthwhile to know about the modification of graphene properties and graphene structural processing effect under the different energy electron beam irradiation.

3.4.1 Experimental Study on the Change of Graphene Structure by Electron Beam Irradiation

The experimental study model is shown in Fig. 3.14. Wherein the graphene is monolayer prepared by the CVD method described previously. The substrate is silicon having oxide with a thickness of 300 nm, and the irradiated electron beam comes from a SEM gun. Firstly, the graphene samples before irradiation were observed in situ and analyzed by Raman spectroscopy, and then the samples were irradiated by electron beam with different duration time, followed by Raman characterization again. In order to prevent the electron beam from rapidly destroying the graphene, the accelerating voltage of the electron beam was 5 kV and the emission current of the electron beam was 0.3 nA. By changing the irradiation time of electron beam, the change of graphene structure under different dose of electron beam was studied. The irradiation time was 0.5, 1, 2, 3, 4, 5, 7, 10 and 15 min, respectively.

Figure 3.15 shows the change of graphene structure under different time electron beam irradiation. Under the action of electron beam, the structures of graphene in irradiated area and unirradiated region differ obviously, which is reflected by the difference in the brightness of the structure (the brightness of the irradiated area is obviously reduced after being irradiated). So it indicates that the electron beam irradiation does have influence on the graphene structure. With the increase of the electron beam irradiation time (2 min), the brightness of the irradiated area continuously reduces, which reflects that higher dose electron beam will have a larger impact on graphene structure. But the folded graphene structure is still clearly seen, so the electron beam at this dose doesn't dispel the graphene structure. When the irradiation time is increased to 10 min, the plicated structure of graphene has

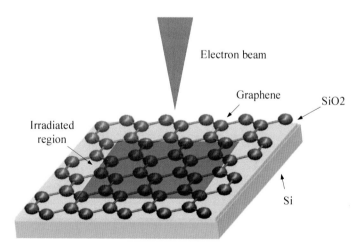

Fig. 3.14 Schematic diagram of the experimental study of graphene structure change under electron beam irradiation

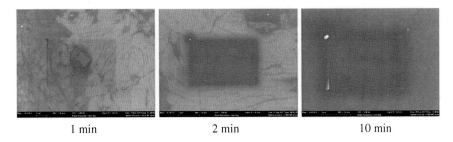

| 1 min | 2 min | 10 min |

Fig. 3.15 Morphological changes of the scanning area under different time electron beam irradiation

become blurred, which shows a very large impact of the graphene structure by electron beam irradiation. If the electron beam acceleration voltage is increased to 30 kV, it is found that it's hard to distinguish the graphene structure under 2 min electron beam irradiation, indicating that graphene structure was wiped away.

Raman spectroscopy signal can well reflect the specific changes of the graphene structure before and after ion beam irradiation. Figure 3.16 shows the Raman spectra of the graphene structure in the scanning region under different time electron beam scanning. Although the original graphene is free from structural defects (Raman signal D peak is almost zero), but under the action of electron beam irradiation, structural defects (D peak) will be generated in graphene very soon (30 s), which corresponds to the change in brightness mentioned above. And as the scanning time increases, the structural defects in the graphene become more and more evident (the intensity of D peak is increasing). From the ratio of the intensity of D peak to G peak, it can be derived that in this period of time graphene gradually turns into nanocrystalline from the complete crystal, which corresponds to continuous decline in the brightness. Within the irradiation time, the graphene structure has not been completely dispelled, and no amorphization has occurred. The results of the above electron microscopy also show that the graphene structure changes

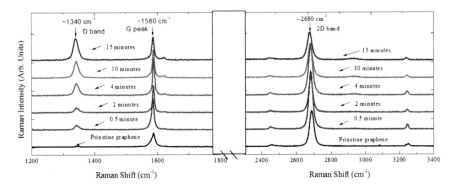

Fig. 3.16 Raman spectroscopic analysis of graphene structure irradiated by electron beam with different duration time

3.4 Interaction Between Electron Beam and Graphene

under irradiation, but its wrinkled morphology is still able to be distinguished. Therefore, at this time the structure of graphene changes to the nanocrystalline under the irradiation of electron beam.

3.4.2 Mechanism of the Destruction of Graphene Structure Under Electron Beam Irradiation

In this section, the MD method was used to simulate the change of graphene structure under electron beam irradiation. The model is shown in Fig. 3.17. The size of the graphene sheet is 10 × 10 nm, with SiO_2 as the supported substrate, and the plane in contact with the graphene is the (0001) plane with the oxygen terminal. In order to reduce the simulation time, the thickness of the substrate is 3.5 nm. The initial distance between graphene and silica is 2.9 Å, and the area of irradiation is 40 × 40 Å. In the process of electron beam action, the bottom atoms of the substrate were fixed. Periodic boundary was applied for the system, and Berebdsen thermal bath was adopted for the atoms near the edge to keep them at 300 K temperature, which can absorb the stress waves generated by the collision.

The treatment of electron beam irradiation is similar to that used in the research of CNTs [12, 13]. Wherein the irradiation process of the electron beam was carried out by periodically applying a certain amount of energy to the PKAs. PKAs were randomly selected from the irradiation area shown in Fig. 3.17, and a single PKA collision occurred every 20 MD steps. This procedure lasted for 2 ps, followed by 1 ps of structural relaxation. Then the next 2 ps collision and 1 ps relaxation was repeated. The whole simulation used a total of 20 times collision-relaxation cycle. Using this method, the temperature of graphene structure is basically below 1000 K, which is in line with the results of the experiment. The dose of the electron beam is expressed as the number of cycles. According to the energy conversion rate calculated in Ref. [14] (the maximum energy absorbed by graphene under 200 keV electron beam is about

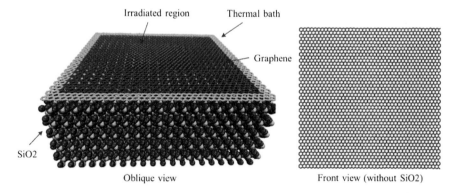

Fig. 3.17 MD model of graphene irradiated by electron beam

43 eV), it is assumed that 0.02% of the incident energy was absorbed by the graphene. Under this energy conversion rate, the changes of graphene structure under electron beam irradiation with energy as 5, 10, 20, 30, 40 and 80 keV were calculated. Meanwhile, in order to compare with the experimental results, the vacancy defects were introduced to the model to analyze the effect of the defects on the impact resistance. Before irradiation, the model was relaxed at room temperature. Then, the graphene was irradiated by electron beams with different parameters with NVE ensemble. After the irradiation, the structure was annealing under 600 K for 100 ps. The temperature was then lowered to 300 K and finally maintained at this temperature for enough time. The LAMMPS software was chosen for simulations. AIREBO potential was used to calculate the interaction between the carbon atoms in graphene, and the interaction between the atoms in SiO_2 was described by the Tersoff potential. In order to describe the cascade collision process, a short-range repulsive force was added to the Tersoff potential. The interaction between the carbon atoms in graphene and the silicon oxygen atoms in the substrate was represented by the L-J potential, where the interaction equation of the L-J potential was expressed as $V_{ij}(r) = 4\varepsilon_{ij}[(\sigma/r)^{12} - (\sigma/r)^6]$, $i = C, j = Si$ or O, r is the interaction distance. The parameters ε_{ij}, σ and the cutoff radius are taken from Ref. [10].

3.4.2.1 Influence of Electron Beam Irradiation on Graphene Structure with Different Irradiation Parameters

Figure 3.18 shows the results of graphene irradiated by electron beam with different parameters. It can be seen that when the electron beam energy is low (5 keV), the graphene structure is not destroyed irrespective of the increase of electron beam dose. At this point, the graphene structure will return to the original morphology after the irradiation of electron beam, which indicates that there is a certain damage energy threshold when the electron beam acts on the graphene. Under the irradiation of high energy (80 keV), the graphene structure will be damaged even under very low-dose electron beam, and there will be also sputtering of atoms and the emergence of vacancy defects in the region close to the destructed area. It indicates that when the energy is relatively large, the lower dose of electron beam irradiation will have a great impact on the graphene structure. While when the electron beam irradiation energy is moderate, the graphene structure will not be destroyed at low doses of electron beam. With the increase of electron dose, the vacancy defects gradually appear in the graphene structure, and then the vacancy defects will increase and fuse, and eventually form a relatively large damage. By analyzing the damage of the graphene structure by different energy electron beam irradiation, the failure threshold of the graphene structure was obtained as 10 keV. According to the results of the simulation, on one hand, we can get the "safe voltage" when using electron beam irradiation to characterize graphene, and on the other hand, the parameters of the electron beam used to process graphene nanostructures can also be obtained.

3.4 Interaction Between Electron Beam and Graphene

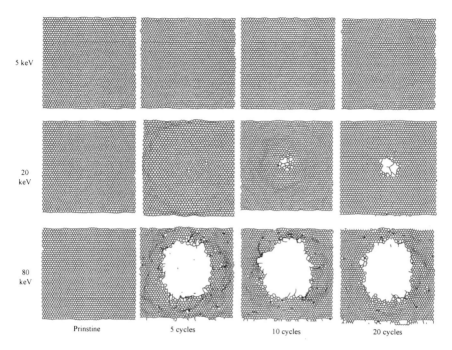

Fig. 3.18 Changes of graphene structure under different doses and energy electron beam irradiation (silica isn't shown)

3.4.2.2 Effect of Vacancy Defects on the Destruction of Graphene Under Electron Beam Irradiation

In the experiment, the graphene structure has obvious structural damage under electron beam irradiation with energy of 5 keV, while the damage threshold from simulation is 10 keV, which is obviously larger than the experimental situation. This phenomenon is due in part to the simplification of the energy conversion of the electron beam irradiating the graphene structure: for the real case, the electron beam irradiation energy is partially absorbed by the carbon atoms in the graphene, and partially absorbed by the substrate, which could reduce the interaction between graphene and the substrate atoms and result in a decrease in the impact resistance of the graphene. While the absorption of the electron energy by the substrate is not taken into account in the simulation. On the other hand, most of the graphene samples used in the experiment are polycrystalline structures, and the graphene structure inevitably has some of vacancy defects, which could reduce its ability to resist particle beam impact. In order to study the effect of defects on the impact resistance of graphene against electron beam, this paper introduced different concentrations of vacancies to investigate the destructive behavior of graphene in the presence of defects.

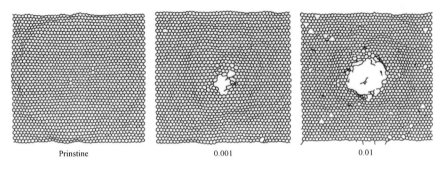

Fig. 3.19 Effect of vacancy defects on irradiation damage of graphene (electron beam energy is 20 keV, the dose is 9 cycles)

Figure 3.19 shows the destruction of the graphene structure by electron beam irradiation in the presence of different concentrations of defects under certain energy and dose conditions, where the defects are randomly distributed in the graphene structure. In the absence of defects, 20 keV, 9 cycles of electron beam irradiation does not substantially destroy the graphene structure (only a small number of polygonal defects), but when a small amount of defects (0.1%) is introduced, there will be obvious structural damage in graphene under electron beam irradiation with the same parameters. When the vacancy defect concentration is relatively large (1%), the degree of damage in graphene structure will be greater. This suggests that the introduction of defects will reduce the ability of the graphene structure to resist particle beam impact, and the degree of reduction will increase with the increase of defect concentration. In experiments, it will inevitably introduce defects to the graphene structure, so we must consider the impact of defect structure during the analysis. In the experiment, there are different forms of defects such as single vacancy, double vacancy, polygon and grain boundary in graphene structure. The existence of these defects can largely decrease the electron beam impact damage threshold of the graphene structure.

Figure 3.20 shows the dynamic process of the variation of graphene structure under impact with different vacancy defect rates. It can be seen that the destruction of the graphene structure experiences three stages: generation of the initial defects, the expansion and fusion of the defects, the stabilization of the defects. The graphene structure with higher vacancy defect rate is more prone to initial defects initiation. Moreover, it can be found that the structural damage of graphene under electron beam irradiation is caused by sputtering of carbon atoms. The sputtered carbon atoms may migrate in the subsequent annealing process and combine with the original vacancy defects in graphene, which could result in the reduction of the vacancy defect concentration in the non-irradiated area after equilibrium of the graphene structure. It also demonstrates that the impurity atoms adsorbed on the surface of graphene are unstable and that they tend to migrate to the locations where the chemical activity is higher in the graphene structure, such as vacancy defects or

3.4 Interaction Between Electron Beam and Graphene

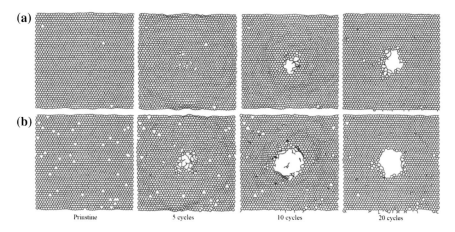

Fig. 3.20 Dynamic change of graphene irradiated by electron beam with **a** 0.1% and **b** 1% vacancy defect rate (electron beam energy is 20 keV)

graphene boundaries. Thus the actual absorbed atoms generated during low energy ion beam implantation are unstable, which will be discussed in Chap. 4.

3.5 Chapter Summary

In this chapter, based on the difference of the energy conversion and structure changes of graphene under irradiated by different species of particle beams, the interaction mechanisms between graphene and laser, ion and electron beam were studied by experiment and simulation. The results were as follows:

1. The damage energy threshold of graphene structure under irradiated by ultrafast laser with different pulse widths was obtained. Under action of ultrafast laser with energy below damage threshold, the in plane expansion and out of plane fluctuation of graphene structure will be intensified. Femtosecond laser can be used to process graphene micro-bands and micron-pores. The damage threshold of multi-layer (3–5 layers) graphene under 50 fs single pulse femtosecond laser irradiation is 0.22 mW. The line damage threshold of graphene structure irradiated by femtosecond laser with a running speed of 0.5 mm/s and pulse frequency of 1000 Hz is 0.11 mW. By controlling the operating parameters, we can achieve the results of graphene thinning and elimination.
2. Under the action of ion beam, the graphene structure will appear four kinds of phenomena with different incident energy: the incident ions are reflected by graphene, the ions are adsorbed on the surface of graphene, the ions are embedded in the graphene structure, and the incident ions transmit through graphene. The presence of the substrate leads to the secondary collision of the graphene structure, which could result in the increase of the damage of graphene

structure, and the secondary collision caused by the sputtering of the substrate atoms is always greater than the direct collision effect from the incident ions.
3. Under electron beam irradiation, the graphene structure will gradually generate some defects, and transform to nanocrystalline structure. When the electron beam irradiation energy is low, the graphene structure will never be destroyed. Under the ideal conditions, the failure threshold of the graphene structure is 10 keV, while the existence of vacancy defects in the actual graphene structure will reduce its failure threshold.

References

1. Wang X, Xu X (2003) Molecular dynamics simulation of thermal and thermomechanical phenomena in picosecond laser material interaction. Int J Heat Mass Transf 46:45–53
2. Ermakov VA, Alaferdov AV, Vaz AR et al (2015) Burning graphene layer-by-layer. Sci Rep 5:11546
3. Roberts A, Cormode D, Reynolds C et al (2011) Respond of graphene to femtosecond high-intensity laser irradiation. Appl Phys Lett 99:051912
4. Wu X, Zhao HY, Zhong ML, Murakawa H, Tsukamoto M (2013) The formation of molecular junctions between graphene sheets. Mater Trans 54:940–946
5. Zakharchenko KV, Katsnelson MI, Fasolino A (2009) Finite temperature lattice properties of graphene beyond the quasiharmonic approximation. Phys Rev Lett 102:046808
6. Lin Z, Ye X, Han J (2015) Precise control of the number of layers of graphene by picosecond laser thinning. Sci Rep 5:11662
7. Lehtinen O, Kotakoski J, Krasheninnikov AV (2011) Cutting and controlled modification of graphene with ion beams. Nanotechnology 22:175306
8. Bellido EP, Seminario JM (2012) Molecular dynamics simulations of ion-bombarded graphene. J Phys Chem C 116:4044–4049
9. Wu X, Zhao HY, Zhong ML, Murakawa H, Tsukamoto M (2014) Molecular dynamics simulation of graphene sheets joining under ion beam irradiation. Carbon 66:31–38
10. Ong Z, Pop E (2010) Molecular dynamics simulation of thermal boundary conductance between carbon nanotubes and SiO_2. Phys Rev B 81:155408
11. Wu X, Zhao HY, Yan D, Pei JY (2015) Investigation of gallium ions impacting monolayer graphene. AIP Adv 5:067171
12. Jang I, Sinnott SB (2004) Molecular dynamics simulation study of carbon nanotube welding under electron beam irradiation. Nano Lett 4:109–114
13. Pregler SK, Sinnott SB (2006) Molecular dynamics simulations of electron and ion beam irradiation of multiwalled carbon nanotubes: the effects on failure by inner tube sliding. Phys Rev B 73:224106
14. Wang H, Wang Q, Cheng Y et al (2012) Doping monolayer graphene with single atom substitutions. Nano Lett 12:141–144

Chapter 4
Doping of Graphene Using Low Energy Ion Beam Irradiation and Its Properties

4.1 Introduction

The doping of graphene has a very important effect on its application. The zero bandgap of the intrinsic graphene limits its application in many fields. The doping method can control its electronic structure and improve the switching ratio of the graphene based transistor to obtain the properties controllable graphene. Meanwhile, doping can modify the chemical activity of graphene surface, making it easier to connect other molecular systems, such as biological aptamers. The conclusion of Chap. 3 shows that when the energy of the ion beam is low, the incident ions can be embedded in the interior of graphene, so as to achieve the effect of doping. In this chapter, the feasibility of graphene doping by low-energy ion implantation was studied by using the experimental method. And then the mechanism of graphene doping by ion implantation was explained by MD simulation method. At the same time, the effect of ion implantation doping on the mechanical and electronic transport properties of graphene was further discussed in order to evaluate the prospect of this method in practical application.

4.2 Experimental Studies of Graphene Doping by Ion Beam Irradiation

4.2.1 Experiment Procedure

The whole experiment is divided into the following parts:

1. Sample preparation

The experimental samples were multilayer graphene (3–5 layers) prepared by CVD method. The preparation of graphene samples was described in Chap. 2. The graphene was transferred to a silicon substrate with a 300 nm oxide layer by chemical etching.

2. Low energy ion implantation

The low energy nitrogen ion beam was implanted into the multi-layer graphene under ambient temperature and pressure conditions. The gas flow rate is 250 mL/min, with upper power as 800 W, and ion beam pulse width as 30 μs. The ion beam energy was controlled in the range of 10–1000 eV, and ion beam dose ranged from 1×10^{14} to 1×10^{15}/cm^2. The failure threshold energy of perfect graphene under the ion beam irradiation is about 20 eV, and the damage threshold of graphene in this paper will be lower than this value due to the existence of a small number of defects. Therefore, in the energy range used in the experimental study, low-energy ion beam irradiation will lead to a certain degree of defects in the graphene structure. For the illustration of the experimental setup, please refer to Ref. [1].

3. Performance characterization

XPS and Raman spectroscopy were used to characterize the graphene performance before and after ion implantation.

4.2.2 Experiment Results of Low Energy Ion Implantation Doping

Figure 4.1 shows the XPS spectra of graphene on a silica substrate. The peak with a binding energy of about 285 eV corresponds to C1s (sp^2-C) in the graphene structure, and the peaks with binding energy of 100 and 532 eV correspond to Si2p and O1s in the substrate silica. When the graphene was not irradiated by ion beam (Fig. 4.1a), there was no peak corresponding to N–C (398.4 eV) binding in the spectrum. After low energy (20 eV) and higher energy (1000 eV) ion beam irradiation, a certain degree of N–C binding peak appears in the graphene structure, indicating that nitrogen ions are embedded into the graphene. However, under different energy ion beam, the peak intensity ratio of the surface material is different, which indicates that different energy ion beam action will have different effects.

Figure 4.2 is the atomic percentage of the chemical elements in the surface before and after ion implantation. The pristine graphene has high carbon element, and no nitrogen. After the ion implantation with different energies, the N element appears. But when the energy of the injected ion beam is high (1000 eV), the carbon element is greatly reduced in the detected material. Which indicates that although the higher energy ion beam will inject the desired ions into the graphene material, the ion implantation process splits the carbon atoms in graphene, causing the graphene material to be largely eliminated. Thus, for the graphene doping by

4.2 Experimental Studies of Graphene Doping by Ion Beam Irradiation

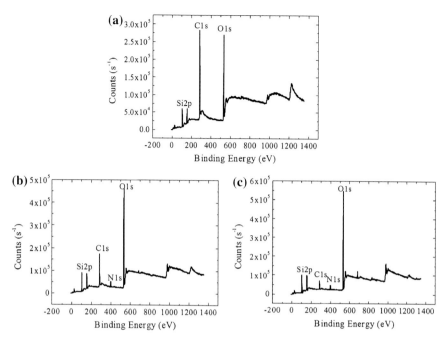

Fig. 4.1 XPS survey spectra of graphene. **a** Pristine graphene, **b** under 20 eV N ion irradiation, **c** under 1000 eV N ion irradiation. The ion dose is $1 \times 10^{14}/cm^2$ for both cases. Reprinted from Ref. [1], Copyright 2017, with permission from Elsevier

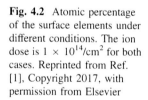

Fig. 4.2 Atomic percentage of the surface elements under different conditions. The ion dose is $1 \times 10^{14}/cm^2$ for both cases. Reprinted from Ref. [1], Copyright 2017, with permission from Elsevier

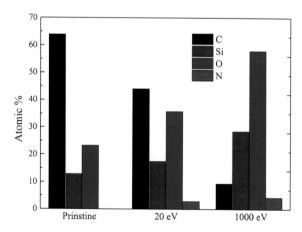

using the low energy ion implantation method, the energy of ion beam must be strictly controlled.

Then the chemical binding state of carbon atoms in graphene was analyzed. The XPS data of pristine graphene, graphene irradiated by 20 eV ion beam and

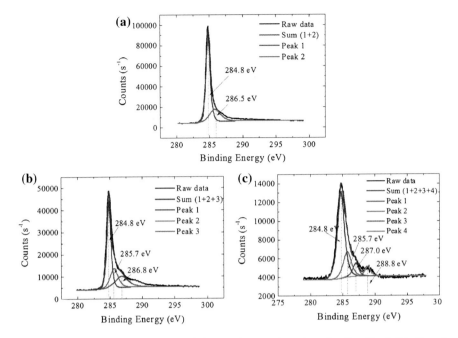

Fig. 4.3 XPS C1s spectrum of the **a** pristine graphene and graphene irradiated by **b** 20 eV, and **c** 1000 eV N ions. The raw data can be split to several Lorentzian peaks at different energy locations. The sum peak is the combination of the split peaks. The irradiated ion doses are both $1 \times 10^{14}/cm^2$. Reprinted from ref. [1], Copyright 2017, with permission from Elsevier

1000 eV ion beam were extracted, and the data were fitted by XPS Peak Processing. The results are shown in Fig. 4.3. The original graphene C1s consist of two parts, corresponding to the sp^2-C bonding state at 284.8 eV and the C–O bonding state at 286.5 eV, respectively. The sp^2-C bonding state is the main bonding form of carbon atoms in graphene, and the C–O bonding state is the bonding form between the carbon atoms in graphene and the oxygen atoms in substrate. Under 20 eV ion beam irradiation, there is almost no change for the sp^2-C bonding state in graphene, and another bonding state at 285.7 eV occurs, which corresponds to the N-sp^2C between the C–N atoms. Meanwhile, the bonding state at 286.8 eV contains the C–O bonding, but it is shifted to the high energy position due to the formation of the N-sp^3C bond between the C–N atoms. Under 1000 eV ion beam irradiation, the surface material still has the sp^2-C (284.8 eV) bonding state and C–O (287 eV) bonding state, and ion beam irradiation will also lead to the formation of N-sp^2C (285.7 eV) and N-sp^3C (287.0 eV) bonding states between carbon atoms in graphene and nitrogen atoms. At this time, there will be some substrate atoms sputtering phenomenon happening due to the high energy ion beam, which will result in the C–Si bonding (288.8 eV) between silicon atoms in substrate and carbon atoms in graphene.

4.2 Experimental Studies of Graphene Doping by Ion Beam Irradiation

In order to understand the existence form of the embedded nitrogen in the graphene structure, the XPS analysis results of the surface nitrogen were extracted and the data were fitted by XPS Peak Processing. The results are shown in Fig. 4.4. Under the 20 eV ion beam implantation, two forms of nitrogen-bonding states appear in the graphene structure, corresponding to substituted nitrogen atom (graphite-N) at 400.1 eV, and pyrrolic-pyridinic-N (pyrrolic-N + pyridinic-N) at 402.0 eV. The substituted nitrogen refers to the N atoms replacing the C atoms during the irradiation. The pyrrolic-pyridinic-N is the N atoms replacing the C atoms in graphene, and meanwhile causing the carbon atoms around to be sputtered. So the pyrrolic-pyridinic-N is actually substituted nitrogen + polygonal vacancy defect. It can be seen that the nitrogen atoms in the graphene are mainly in the form of replacement under the 20 eV ion beam irradiation. When the ion beam energy increases (1000 eV), besides the substituted nitrogen at 399.7 eV and pyrrolic-pyridinic nitrogen at 401 eV, there is also a significant peak at 398.3 eV, where the N and Si atoms are bonded, which is formed mainly due to the sputtering of substrate atoms under the higher energy ion beam irradiation, causing the combination of nitrogen atoms and the base atoms. The positions of each peak shift to lower energy, this is because that there are more sp-C and sp^3-C bonds generated under the higher energy ion beam irradiation, and their combination with the nitrogen atoms will produce the shift of the peak position.

Figure 4.5 shows the results of the Raman spectroscopy analysis of the graphene structure before and after irradiation. Before the ion beam irradiation, there is a small amount of defects (D peaks) in the graphene structure, which is unavoidable in the preparation and transfer of graphene. After the ion beam irradiation, the defective peak in the graphene structure increases, which indicates that the implantation of low energy ion beam into graphene can cause the destruction of the graphene structure. When the dose or energy of the ion beam increases, the defects in the graphene will increase obviously. If the energy of the injected ion beam is particularly high (1000 eV), the incident ion beam will substantially completely

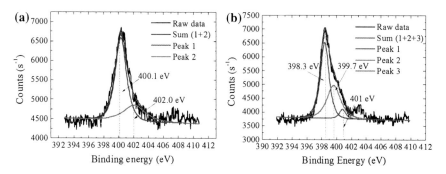

Fig. 4.4 XPS N1s spectrum of the graphene irradiated by **a** 20 eV, and **c** 1000 eV N ions. The raw data can be split to several Lorentzian peaks at different energy locations. The sum peak is the combination of the split peaks. The irradiated ion doses are both $1 \times 10^{14}/cm^2$. Reprinted from Ref. [1], Copyright 2017, with permission from Elsevier

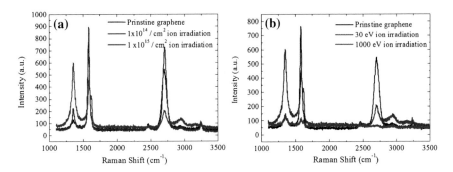

Fig. 4.5 Raman spectra of the graphene irradiated by N ions with **a** different ion dose, ion energy is 20 eV, and **b** different ion energies, ion dose is $1 \times 10^{14}/cm^2$. Reprinted from Ref. [1], Copyright 2017, with permission from Elsevier

destroy the graphene structure (disappearance of the graphene characteristic peaks). Figure 4.2 also shows that most of the carbon atoms in the surface material are eliminated at higher energy ion beam implantation. When doping the graphene structure using low energy ion implantation, it is desired that more ions are implanted into the graphene structure, and on the other hand, it is also required to reduce the defects caused by ion beam implantation as much as possible, which is closely related to the parameters of the ion implantation process. Therefore, the energy and dose of the injected ion beam must be strictly controlled for ion beam implantation doping, which will be studied later using MD simulations.

4.2.3 Summary of the Experiment

The method of low energy ion implantation can be used to dope graphene: on the one hand, the implantation of ions can dope graphene by means of displacement, and on the other hand, the implantation doping is accompanied by the vacancy defects. The higher energy ion beam will mainly induce the defects into graphene, and the increase of the ionic dose will lead to the further destruction of the graphene structure. Therefore, it is needed to select the appropriate ion beam energy and dose for the doping of graphene by low-energy ion beam irradiation.

4.3 Theoretical Analysis of the Doping Mechanism

In this section, the doping mechanism of graphene by low energy ion implantation was analyzed by MD simulation and first-principle calculation methods. The influence of the ion beam energy and dose was discussed.

4.3.1 Research Model

Nitrogen ion was selected as implanted ion to be the same as experiment, and the simulation does not take into account the charge of ions. The initial position of the nitrogen ion is 60 Å from the graphene plane and was then randomly injected into the graphene sheet in the direction perpendicular to the graphene plane. The area of the irradiated region is 40 × 40 Å. The graphene sheet has a size of 100 × 100 Å, and the edge (green region) was fixed. The near edge region (red region) uses a thermal bath at 300 K to absorb the stress wave generated by ion beam shock. The interaction between carbon atoms was described by the AIREBO potential function. The interaction between the incident nitrogen ions and the carbon atoms in the graphene was described by the Tersoff potential function. An exclusion term was added to the Tersoff potential function in the short-range distance to describe the exclusion interaction between graphene and nitrogen ions. The classical MD simulation was carried out using LAMMPS, and NPT ensemble was used for the ion implantation. The energy of the ion beam was defined as the kinetic energy applied to the ion, which ranges from 10 to 100 eV. The dose of the ion beam was defined as the number of ions incident to the irradiation area, ranging from 20 to 320. The detail illustration of the model can be found in Ref. [1].

4.3.2 Variation of Graphene Structure Under Nitrogen Ion Implantation

Figure 4.6 shows the structure of graphene under different parameters of ion beam irradiation. It can be seen that the graphene structure will change under ion beam irradiation with different parameters. After ion beam irradiation of graphene, some of the ions will be adsorbed in the graphene surface in the energy and dosage range used. Since the energy of the incident ions is greater than the failure threshold of graphene structure, the carbon atoms in the graphene are partially sputtered to form vacancies. A small amount of incident ions will replace carbon atoms in the graphene structure to form replacement dopants. Besides, the energy and dose of the incident ion beam will affect the change of graphene structure: when the energy of the incident ion beam is low, there is not enough energy to make the carbon atoms in the graphene structure sputtering, so that most of the incident ions will exist in the form of adsorption. With the increase of ion beam energy, more and more carbon atoms in graphene will be sputtered, which could lead to more vacancy defects, and more incident ions will replace the carbon atoms to generate substitution doping. When the energy or dose of the incident ion beam is too high, large number of carbon atoms will be sputtered from graphene structure, resulting in serious structural damage.

After the implantation of ion beam, there will be large number of adsorption atoms and vacancy defects in the graphene structure, will these adsorption atoms

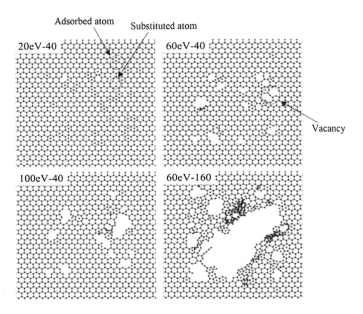

Fig. 4.6 Structures of graphene under ion irradiation with different parameters. The blue balls represent carbon atoms, and the red balls represent implanted N ions. Ion dose is described by the number of implanted ions. Reprinted from Ref. [1], Copyright 2017, with permission from Elsevier

and vacancy defects be stable for long time? The atoms adsorbed on the graphene surface are unstable, and the adsorbed atoms easily immigrate under the energy minimization of the system. The defect structure in graphene is also energetic and unstable. It is easy for the defects to combine with impurity atoms from the surrounding environment to reduce the energy of the system. The previous results of defective graphene under electron beam irradiation also shows that the sputtered carbon atoms can remedy the original defects in graphene. So what is the steady-state doping structure of graphene under ion implantation conditions? At the same time, because the energy and dose of the incident ion beam have a great influence on the doping structure of graphene, when the energy or dose of the ion beam is too large, there will be non-remediable defects generated, which would inevitably arouse the discussion of the influence of ion beam irradiation parameters.

4.3.3 The Energy of the System Corresponding to the Different Doping Configurations

So the low energy ion beam incident will firstly induce the formation of adsorbed ions and vacancy defects. While these adsorption defects were not found in the experiment. In order to explore the steady-state structure of the graphene after irradiation, the

4.3 Theoretical Analysis of the Doping Mechanism

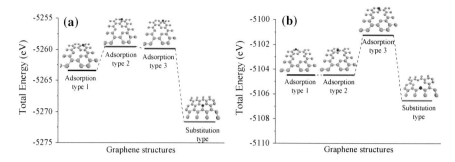

Fig. 4.7 Stable energy of the structures generated by ion beam irradiation. **a** The case of formation of monovacancy, and **b** the case of formation of bivacancy. Reprinted from ref. [1], Copyright 2017, with permission from Elsevier

change of graphene structure under the existence of adsorption atoms and vacancy defects were discussed. In this thesis, several different structural forms were optimized by SIESTA ab initio package and the minimum energy of the systems was calculated. Figure 4.7 shows the steady state energy of graphene structure with different types of defects. The structures with the adsorbed ions right above the carbon atom (adsorption type 3), on the top of carbon-carbon bond (adsorption type 2), and on the top of hexagon structure center (adsorption type 1) were considered as typical adsorption state. It is found that regardless of the position of the adsorbed atoms, the energy of the substitution type system is always lower than the energy of the adsorption type system. For the structures of bivacancy defect + adsorbed ion, the same conclusion can be drawn, i.e., the energy of adsorption type system is always greater than the substitution type system. Therefore, the adsorbed ions generated in Fig. 4.6 are not stable and they are easy to migrate to the vacancy sites in the graphene structure. When the number of vacancy defects and the number of adsorbed atoms in the graphene structure are comparable, most of the adsorbed atoms will gradually migrate to the vacancy defects position to reduce the energy of the system, resulting in the substitutional doping or substitution doping + vacancy defects, which is also consistent with the experiment results. We also found the phenomenon that the adsorption atoms could remedy the vacancy defects in Chap. 3 when studying the electron beam irradiation with defective graphene. When the number of vacancy defects in the graphene structure is small, the redundant adsorbed atoms on the graphene surface will eventually migrate to the edge of the structure (the energy at the edge of the system tends to be higher).

Figure 4.8 shows the transformation of the several different defective structures formed by ion beam irradiation during the optimization process. For the structures of the adsorbed atoms + vacancy defects formed in Fig. 4.8, they easily form substitution doping (when the number of adsorbed atoms equals the number of sputtered carbon atoms) or the form of substitution doping + polygonal defects (when the number of adsorbed atoms dose not equal the number of sputtered carbon

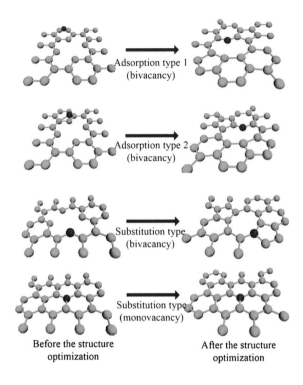

Fig. 4.8 Variation of the structures of graphene with different types of defects before and after structure optimization. Reprinted from Ref. [1], Copyright 2017, with permission from Elsevier

atoms) in the optimization process. The substitution nitrogen atom states (399.7 eV) and the pyrrole-pyridine nitrogen atom states (401 eV) observed in the experiments correspond to the two different forms of doping states. The doping structure of graphene in the nitrogen atmosphere was also reported in [2, 3], which is consistent with the results of this study.

Therefore, the doping form and doping mechanism of the graphene structure under low-energy ion implantation are as follows: after the low energy ion implantation, the adsorbed atoms and vacancy defects are formed in the graphene structure at first. In the subsequent steady-state equilibrium process, the adsorbed atoms will migrate to combine with the vacancy defects, resulting in substitution doping. When the defects formed by irradiation in graphene are complicated (double vacancies or multiple vacancies defects), the defects cannot be completely remedied, resulting in the formation of substitution doping + polygonal defects. In Ref. [4], it was mentioned that a two-step method can be used for the chemical doping of the graphene structure, that is, high-energy particles are used to bombard the graphene to obtain a structure with defects and then deposit the target ions into the graphene to form doped structure. Due to the poor chemical activity of the C–C bond in perfect graphene, it is difficult to form a direct doping, while this method can overcome the above difficulty. The doping mechanism of low-energy ion implantation is similar as that in Ref. [4], nevertheless, it can achieve the doping results in one step, which has obvious advantages.

4.3.4 Influence of the Energy and Dose of Implanted Ion Beam

For the doping process, it is needed to reduce the introduction of defects, meanwhile, it must ensure that there will be enough ions implanted into graphene structure at a certain dopant dose. It has been confirmed that the doping process of graphene structure is divided into two steps: firstly the formation of absorbed ions and vacancy defects, and then migration of adsorbed atoms and remediation of the vacancy defects in the structure. Therefore, in order to implant enough ions into the graphene structure, the number of adsorbed atoms cannot be excessive, too many adsorbed atoms will migrate to the edge of graphene, which will reduce the doping efficiency. Meanwhile the number of vacancy defects (sputtered carbon atoms) cannot be too excessive, otherwise the generated doping structure will be always accompanied with vacancy defects. Therefore, it is necessary to control the energy and dose of the ion beam, so that the number of sputtered carbon atoms and the amount of adsorbed nitrogen ions are comparable, which could make it possible to obtain a good doping structure.

Figure 4.9 shows the number of adsorbed ions and sputtered carbon atoms in the graphene structure under different energy nitrogen ion irradiation. With the increase of ion energy, the collision between atoms will become more and more intense, so the number of sputtered carbon atoms in the graphene structure will gradually increase, and the number of nitrogen ions embedded in the graphene structure will gradually decrease. When the energy of the incident ions is about 65 eV, the number of adsorbed nitrogen ions and the number of sputtered carbon atoms are almost the same, and the defects in the graphene are basically single and double vacancy defects, so during the equilibrium process, the adsorbed nitrogen ions will fill the formed vacancy defects well. Figure 4.9b shows the change in the ratio of adsorbed atoms to sputtered atoms. Their ratio is close to 1 in the vicinity of 65 eV, where the best doping effect can be achieved. Figure 4.10 shows the effect of ion

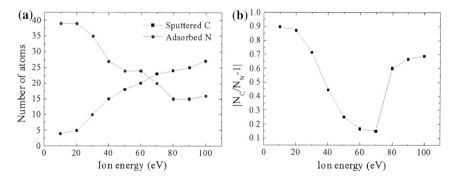

Fig. 4.9 Influence of irradiated ion energy on the doping results, the ion dose is set as 40. **a** Variation of the number of sputtered C atoms and adsorbed N ions. **b** Variation of the ratio between the number of adsorbed atoms and sputtered atoms. Reprinted from ref. [1], Copyright 2017, with permission from Elsevier

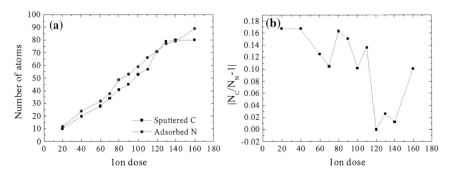

Fig. 4.10 Influence of irradiated ion dose on the doping results, the ion energy is set as 60 eV. **a** Variation of the number of sputtered C atoms and adsorbed N atoms. **b** Variation of the ratio between the number of adsorbed atoms and sputtered atoms. Reprinted from Ref. [1], Copyright 2017, with permission from Elsevier

dose on the number of adsorbed nitrogen atoms and sputtered carbon atoms. It can be seen from Fig. 4.10a that when the ion beam energy is 60 eV, the number of sputtered carbon atoms is basically the same as the number of adsorbed nitrogen atoms with different dose ion beam irradiation. Figure 4.10b also depicts that the ratio between the number of adsorbed atoms and the number of sputtered atoms is always close to 1, so at this time the defects structure is always likely to be well remedied by the adsorbed atoms in the subsequent equilibrium process. However, as shown in Fig. 4.6, when the dose of ion beam is too large, the graphene structure will be seriously damaged, resulting in a defective structure hardly to be filled by nitrogen atoms, which means it is impossible to get the desired substitution doping structure. Thus, the dose of ion beam cannot be too large. By analyzing the doping structures of graphene under different ion dose, the optimal ion dose is found to be 50 (corresponds to 3.125×10^{14} cm^{-2}). Under this value of ion dose, the induced defects are mainly simple vacancies (monovacancies, bivacancies or trivacancies), which are easily filled up by the migration of adsorption atoms to form a good substitution doping result. Of course, any ion dose below this value can also lead to a good substation doping result, but the efficiency will be lower.

In summary, under the low energy nitrogen ion beam implantation, the optimum ion energy for graphene structure doping is 65 eV and the optimal ion dose is 50 (3.125×10^{14} cm^{-2}). These results need to be further confirmed in the experiment.

4.4 Mechanical Properties of Doped Graphene by Ion Beam Irradiation

The mechanical properties of graphene are excellent, but whether the graphene can keep its good mechanical properties when doped by ion implantation? This section discussed the mechanical properties of graphene after doping, which is significant for the application of the low energy ion implantation doping method.

4.4 Mechanical Properties of Doped Graphene by Ion Beam Irradiation

4.4.1 MD Simulation Model

Figure 4.11 shows the research model of mechanical properties. It has been previously shown that the final result of low energy ion implantation doping is the substitution doping or substitution doping + vacancy defects, so only the two forms of doped structure were discussed. Wherein, for the doping structure without forming vacancies, it is further divided into uniform doping and selected doping (doping atoms are localized). For the doping structure with vacancy defects, besides the substitution doped structure, some concentration of vacancy defects (delete a certain percentage of carbon atoms) were randomly introduced into graphene structure. Then the influence of defects introduced by ion beam implantation on the mechanical properties of graphene was discussed. In addition, the effects of different doping concentrations and different doping atomic types were also discussed. The size of the graphene sheet was 20 × 20 nm, with the left side carbon atoms being fixed, and the right edge carbon atoms being stretched with a speed of 0.2 Å/ps along the horizontal right direction. The tensile load was along the armchair-type

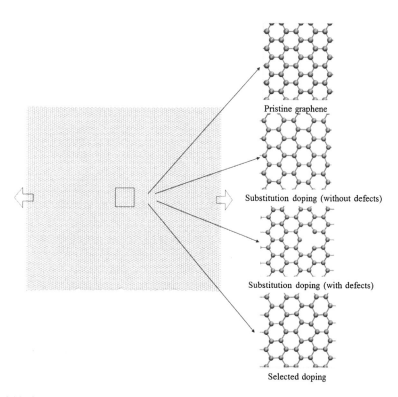

Fig. 4.11 Study model of mechanical properties of doped graphene (the right image shows the enlarged image of the middle region on the left. The green balls represent the dopant atoms, which are distributed in the whole graphene plane)

edge, and the model used a free boundary condition. The interaction between carbon and carbon atoms in graphene was described by AIREBO. The cutoff distance of C–C bond was 1.92 Å. The interaction between carbon atom and dopant atoms was described by Tersoff potential. The simulation software was LAMMPS. Before the uniaxial tension, the system was fully relaxed under the condition of 300 K. The stress and strain extraction method of the system after stretching was described in Chap. 2.

4.4.2 Effect of Implantation Doping on Tensile Stress Distribution of Graphene

Since the size of the dopant atoms is different from that of the carbon atoms, and the bonding force between the dopant atoms and the carbon atoms is different from that of the carbon atoms in the graphene, so the mechanical properties will be different after the element is doped into the graphene structure, and the existence of the vacancy defect will further affect the mechanical behavior. Figure 4.12 shows the calculated atomic stress distribution of the original graphene, substituted doping graphene (with a concentration of 12.5%), and substituted doping + vacancy defect graphene under uniaxial tensile loading. For the original graphene, due to the uniform crystal structure, there is no obvious stress concentration in the graphene plane. While the stress distribution of the graphene structure changes obviously when the low energy ion was implanted into the graphene structure as substitution doping. There is a significant stress concentration near the nitrogen atom. The uneven distribution of the stress could lead to uneven deformation and failure during the stretching process. When the implanted ion beam introduces part of the vacancy defects while forming the substitution doping, the presence of the defective state will further change the distribution of the stress in the graphene plane, and there is still a stress concentration near the nitrogen atom. However, the atomic force in the graphene plane is significantly lower than that of the original graphene at the same tensile strain.

For the tensile stress-strain relationship of the doped structure, the effect of displacement doping on the tensile strength of graphene structure is small, but it will reduce the fracture strain. When the substitution doping and vacancy defects exist simultaneously, both of the tensile strength and fracture strain of graphene structure will greatly reduce. Meanwhile, ion implantation doping will also reduce the elastic modulus of graphene structure. Since the amount of vacancy defects introduced by the implantation doping is closely related to the parameters of the ion beam, it is necessary to strictly control the parameters of the implanted ions.

4.4 Mechanical Properties of Doped Graphene by Ion Beam Irradiation

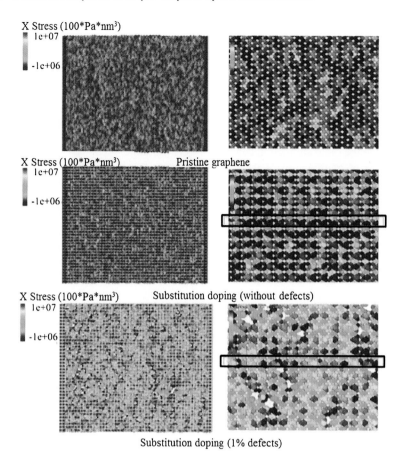

Fig. 4.12 Effect of implanted ion doping on the stress distribution of graphene (taken at 0.12 tensile strain). The black wireframe represents the entire row of doping. The dopant atoms are amplified to distinguish the carbon atoms, and the right image is the magnified display of the left

4.4.3 The Influence of Doping Concentration and Doping Ion Distribution

In the actual doping process, it is often required to dope graphene with different concentrations and different distribution (selected doping) conditions. The stress distribution of the graphene structure may be different under doping with different concentration and different distribution. Figure 4.13 shows the tensile mechanical properties of the graphene structures under two doping concentrations. It can be seen from Fig. 4.13a that the stress distribution of the graphene structure is close to that of the original graphene structure under low nitrogen concentration doping, and no obvious stress concentration occurs. When the concentration of element is increased, there will be obvious stress concentration gradually appeared in the graphene structure, which will affect its tensile damage. It can be seen from Fig. 4.13b that the stress-strain relationship of the graphene structure in the low concentration (3.125%) doping is close to that of the original graphene. In the case of high concentration (12.5%) doping, the fracture strain will be reduced, while the tensile strength will almost keep unchanged. The elastic modulus of the doped graphene structure will decrease with both doping concentrations.

Figure 4.14 shows the tensile stress-strain distribution of doped graphene with different dopant ions distribution. Three different aggregation cases were discussed here, while the other simulation conditions were the same. It can be clearly seen that the degree of aggregation of the dopant elements has a significant effect on the tensile mechanical properties of graphene. When dopant atoms are more localized, the tensile strength and the fracture strain of graphene are obviously decreased, and the elastic modulus of the structure decreases. From the previous results, it is known that the substitution doping can cause obvious stress concentration in the vicinity of the embedded atoms. When the dopant atoms are more localized, the stress concentration of graphene structure becomes more obvious, which would lead to the occurrence of different doping properties. In the actual experiment, it is often necessary to carry out the selected doping, which can cause the large stress concentration of graphene in local region, so special attention is needed for this case. Meanwhile, this paper presents the dynamic variation process of graphene structure under tension in the case of high degree of atoms aggregation (type 3), as shown in Fig. 4.15. Due to the stress concentration in the doping site, it will be easier for the crack to initiate at the localized doping site during stretching, and then the crack will spread along the direction perpendicular to the stretching. The expended crack will coalesce with the cracks initiated from other doping sites, resulting in the overall destruction of the graphene structure. The time from crack initiation to propagated fracture is very short. At this time, the fracture strain of graphene is around 0.117, which is much lower than the value of pristine graphene and uniform doping condition.

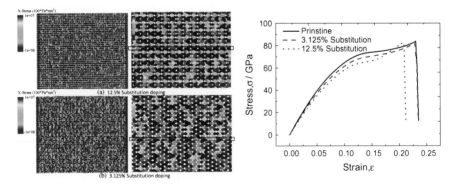

Fig. 4.13 **a** Atomic stress distribution and **b** tensile stress-strain relationship of the graphene structures doped with different concentrations

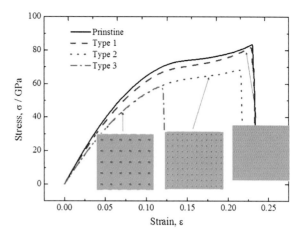

Fig. 4.14 Tensile stress-strain curves of doped graphene with different dopant distributions (the total number of doped atoms is the same for different types)

4.4.4 Influence of Defect Concentration and Doping Element Type

The defects introduced by the doping process will reduce the mechanical properties of graphene, and different concentrations of defects will have different influence on the tensile mechanical behavior. Figure 4.16 shows the dynamic changes of the stress distribution of doped graphene structures with different concentrations of defects. When the defect concentration is low (0.1%), the graphene structure has a relatively large failure strain. At the strain of 0.162, the graphene structure begins to generate crack and the crack spreads along the path perpendicular to the stretching direction. With the increase of defect concentration, the location and time of the initiation of crack in graphene will change, for example, at 0.5% concentration of defects, the crack initiate at 0.152 strain on the right edge. When the defect concentration is high (1%), the strain at which the crack initiates will be greatly reduced (0.123). At this time, the graphene structure is more prone to be fractured.

In addition to the above factors, the bonding force between the dopant atoms and the carbon atoms in the graphene structure will be different if different dopant elements are adopted, which could result in a change in tensile mechanical behavior. Figure 4.17 shows the tensile mechanical behavior of the graphene structure under substitution doping of N element and B element. It can be seen from Fig. 4.17a that nitrogen doping can cause obvious stress concentration in the doping site, but there is no obvious stress concentration in the graphene structure after boron doping. This shows that the bonding force between boron and carbon atoms is close to that of carbon atoms in graphene, and there is a relatively uniform force in the process of stretching. It can also be seen from the corresponding tensile stress-strain relationship in Fig. 4.17b that the stress-strain relationship of the boron-doped graphene structure is close to that of the original graphene, but the doping will always cause elastic modulus of the graphene structure to decline.

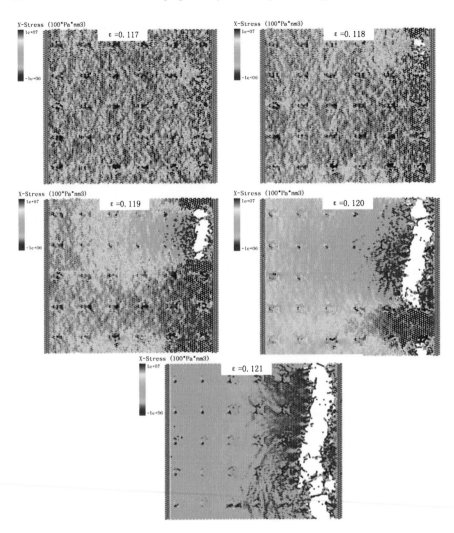

Fig. 4.15 Dynamic change of the tensile stress distribution of graphene under type 3 doping condition

4.5 Electronic Transport Properties of Doped Graphene

After doping of graphene, in order to explore its application in electronic devices, we must also study its electrical properties. In this section, the DFT and NEGF methods were combined to calculate the electronic transport properties of doped graphene, taking into account the influence of different doping forms and doping positions on the electronic transport properties.

4.5 Electronic Transport Properties of Doped Graphene

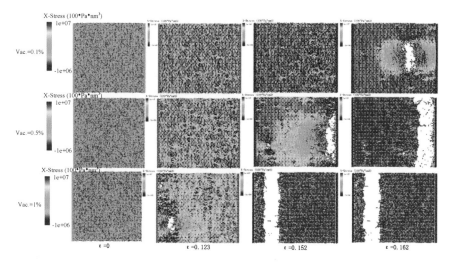

Fig. 4.16 Dynamic changes of the tensile stress distribution of doped graphene with different concentration defects

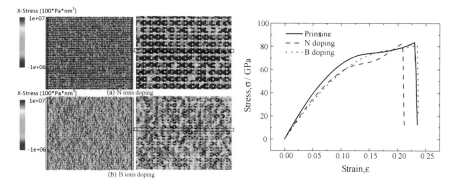

Fig. 4.17 a Atomic stress distribution and **b** tensile stress-strain curves of graphene when substitutionally doped by N and B elements

4.5.1 Research Model

Figure 4.18 shows the study model of the electronic transport properties of doped graphene. ZGNR with the width N = 10 (N is the number of zigzag carbon chains) was used for all simulations. The shaded portions represent the left and right electrodes, which are essentially semi-infinite graphene nanoribbons, and the middle region is the scattering area. To build a model with substitution doping, the carbon atoms in the scattering region were replaced with substitutional dopants, and at the same time, a single (double) vacancy defect was formed by removing one or two carbon atoms in the vicinity of the replaced atoms to simulate the formation of

Fig. 4.18 Study model of electronic transport properties of doped graphene

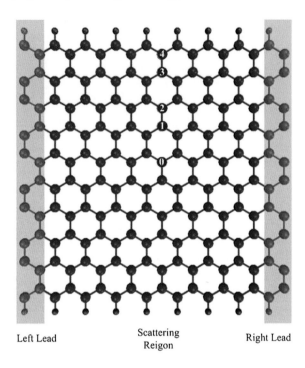

Left Lead   Scattering Reigon   Right Lead

defects during the substitution doping. The numbers in Fig. 4.18 indicate the positions occupied by the dopant atoms. Number 0 indicates the position at which the atoms are replaced when studying the influence of the dopant forms and the dopant element species, and the positions 1, 2, 3, and 4, which are perpendicular to the transport direction, are used to study the influence of the location of the dopant atoms on the electronic transport properties. The upper and lower edges of the model were saturated by adsorption hydrogenation. A 15 Å vacuum layer was added at both ends of the boundary with hydrogen atoms.

In this paper, the geometrical optimization, the electronic transmission coefficient, and the current and voltage characteristics were calculated using the Transiesta software package described in Chap. 2. The GGA was used as the exchange correlation function. The cutoff energy of the plane wave was 150 Ry, and the K point grid of the Brillouin zone was $1 \times 1 \times 100$ (where Z is the electronic transport direction). The calculation does not take into account the effects of spin.

4.5.2 Electronic Transport Properties Under Different Doping Forms

Figure 4.19 shows the results of the transport lines in different doping forms. The distribution of the transmission function of the original graphene nanoribbon is

4.5 Electronic Transport Properties of Doped Graphene

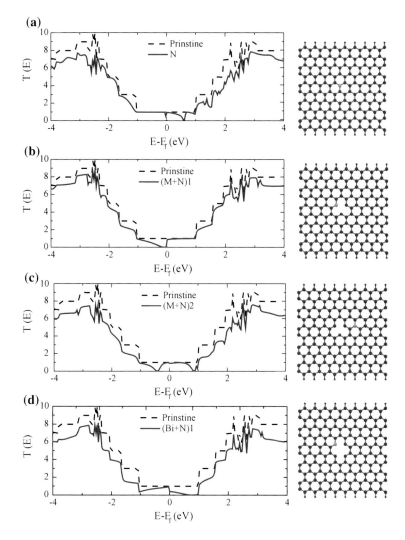

Fig. 4.19 Electronic transport lines (zero bias) of graphene nanoribbon under different doping forms. The doping forms with **a** substitution (N), **b** substitution + single vacancy [(M + N)1], **c** substitution + single vacancy [(M + N)2], and **d** substitution + double vacancy [(Bi + N)1] were considered. The image in the right shows the corresponding doping structure. Dashed line is the results of pristine graphene

consistent with those reported in Ref. [5, 6]. When different forms of doping structure are introduced, the quantum step characteristic of the graphene transport line disappears, and the carrier transmission probability of the graphene nanoribbon structure is inhibited by the existence of the dopant structure. At some energy level, the transmission capacity is zero. As can be seen from Fig. 4.19a, the substitutional doping of nitrogen ions causes the valley with the Lorentz shape at the energy above the Fermi level (about 0.6 to 0.7 eV), and the minimum transport coefficient corresponding to the valley position is zero, which means that at the energy $E - E_f = 0.65$ eV, the transport will be completely reflected, and the transmission is blocked by impurities, causing the zigzag graphene nanoribbon to transform from metal to semiconductor type. While the transport line on the other side near the Fermi level is consistent with the original graphene, which means this doping type forms the donor semiconductor. When doping process induces a single vacancy defect into the graphene structure, the relative position of the defect and the dopant atom will have an effect on the transport performance: when the dopant atoms are at the edge of the vacancy defect (Fig. 4.20b), there will be a valley at the energy below Femi level (at around −0.05 to −0.1 eV). The lowest transport coefficient is zero at this valley, that is, when the electron energy is about $E - E_f = -0.1$ eV, the transport will be completely reflected. While the transport line in the vicinity of energy above the Fermi level is consistent with the original graphene, which means at this doping form the graphene nanoribbon is converted from the donor semiconductor to the acceptor semiconductor (this can also be obtained from the band structure). This transition is caused by the coupling between the dopant atoms and the vacancy defects local state, and the acceptor doping caused by the vacancy defect dominates. When the dopant atoms are separated from the vacancy defects (Fig. 4.19c), the coupling between the atomic states caused by the dopant atoms and the vacancy defects is weak, and there will be transmission valley appearing at the energy both below and above the Fermi level. When the electron energy is $E - E_f = 0.8$ eV and $E - E_f = -0.5$ eV, the transport will be completely reflected. At this time, the semiconductor structure formed by the doping has the characteristics of vacancy defects and substitution doping. While in the vicinity of the Fermi level, the transport line of the doped structure is consistent with the original graphene. When the doping process introduces double vacancy defect into graphene (Fig. 4.19d), there will be a greater degree of suppression for the transport properties at the energy close to Fermi level. There is a wide region with zero transmission coefficient at the energy above the Fermi level (0.2–1 eV), which indicates that the presence of double vacancy defects could cause an increase in the energy gap of the graphene structure, making it a semiconductor with greater bandgap width.

Figure 4.20 shows the current-voltage curves of the graphene nanoribbon structure under different doping forms. It can be seen from the figure that the current of the original graphene nanoribbon under the bias voltage is very small due to its symmetrical structure, which would induce the increased zero transmission band gap under the increased bias [6]. However, with the introduction of dopant elements, the symmetry of the graphene structure is destroyed, and the electronic structure near the Fermi level has changed significantly, so that the transmission line of graphene under the bias voltage changes, which could result in the increase of the current of the structure. When the doping process introduced defects into the graphene nanoribbon structure, the defects will further destroy the structural symmetry of the original graphene, but at the same time the defects will bring a new local state, which will suppress the transport capacity of the graphene structure.

4.5 Electronic Transport Properties of Doped Graphene

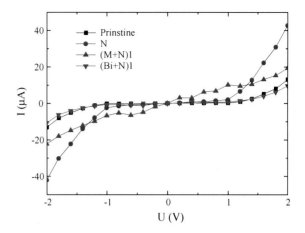

Fig. 4.20 Current-voltage relationship of graphene nanoribbon under different doping forms

At this time, the local state states caused by the defect state and the doping element are coupled to produce a more complex effect: when the defect is in the form of a single vacancy, it will cause the structural current to increase at low bias voltage (<1.2 V) (compared to substitution doping), and the structural current to decrease at high bias voltage (>1.2 V). When the defect exists in the form of double vacancy, the local state of the defect produces a strong suppression on the structural transport capacity, resulting in a small current through the structure. At this time, the current through the graphene structure is substantially the same as that of the original graphene.

Since the doping forms formed in the graphene are closely related to the parameters of the implanted ions, it is possible to obtain different forms of graphene dopants by controlling the energy and dose of the implanted ions, which will provide theoretical support for the design of the nanoelectronic device.

4.5.3 Effect of Doping Position on Electrical Performance

One of the advantages of low-energy ion implantation doping of graphene structure is the ability to do the doping at specified location with specified concentration. In order to study the influence of doping location on the transport properties of graphene nanoribbon, this section used different position doping of graphene described in Fig. 4.18 to calculate the transport lines at zero bias.

It can be seen from Fig. 4.21 that the transport capacity of the graphene nanoribbon structure is inhibited at different doping sites, and the closure of the transport channel (i.e., zero transport coefficient) occurs near the Fermi level. This is due to the fact that doping leads to changes in the electronic structure near the Fermi level, resulting in a localized state structure, and leading to the transport of electrons being reflected. When the position of the dopant atom is in the position

Fig. 4.21 Electronic transport lines of graphene nanoribbon doped at different locations (zero bias)

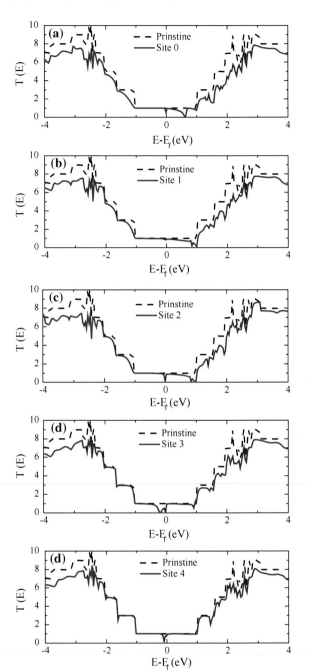

near the center of the graphene nanoribbon (site 0, site 1, site 2), the doping element has a strong inhibitory effect on the transport capacity, and the generated transmission valleys are all at energy above Fermi level. At this time, the zigzag graphene nanoribbon exhibits the characteristic of the donor semiconductor, and the position of the transmission valley will be shifted toward the higher energy position as the impurity atom moves closer to the edge. With the change of the location of the doping atom, the inhibitory effect of the dopant elements on the transport capacity will get weaker when they are at the nanoribbon edge position (site 3, site 4). At this case, the electronic transport capacity is close to the original graphene nanoribbon at different energies. Especially when the dopant element is in the site 4 position, the transport line is almost the same as the original graphene. In addition, when the dopant element is near the edge position of the graphene, its transmission cut-off position is shifted from the position above the Fermi level to the position below the Fermi level, which means that the graphene nanoribbon may be gradually converted from the donor semiconductor to acceptor semiconductor features. It is also noted in the literature [7] that the change in the doping position of the B element can cause a change in the semiconductor characteristics. In general, with the position of the dopant element gradually closing to the edge of the graphene nanoribbon, the transport capacity of the graphene structure at zero bias gradually approaches the original graphene. This is because the transport capacity of the graphene nanoribbon is determined by the local state caused by both the edge state and the impurity element. When the impurity element is far from the edge position, the coupling effect between the impurity state and the edge state is weak, and the impurity element will have larger effect on the transport capacity of the graphene structure. When the impurity element is in the edge position of the graphene structure, the impurity state and the edge state will be obviously coupled. The influence of the impurity element on the transport capacity of the graphene structure will be greatly reduced. Therefore, the transport properties of graphene can be adjusted by controlling the position of the dopant element in the graphene structure.

4.6 Summary

In this chapter, the graphene structure was doped by low energy ion implantation, and the phenomena and mechanism of doping were analyzed by experiment and simulations. The mechanical and electronic transport properties of the doped structure were discussed. The results are as follows:

1. Low energy ion beam implantation can be used to dope graphene. On the one hand, the ions will be doped into graphene structure in the way of substitution, on the other hand, there will be some vacancy defects introduced during the doping process. The defects will dominate during the high energy ion beam irradiation. And the increase of ion dose will lead to further destruction of the graphene structure.

2. After low-energy ion implantation, firstly there will be adsorption atoms and vacancy defects generated in graphene. During the subsequent equilibrium process, the adsorbed atoms will combine with the vacancy defects to form replacement dopants. When the defect form is complicated, the defects cannot be completely remedied, which would lead to the formation of doping structure as substitution doping + polygon defects. Under the condition of low energy nitrogen ion beam implantation, the optimum energy of graphene structure doping is 65 eV and the optimal dose is 3.125×10^{14} cm^{-2}.
3. Under the condition of higher concentration doping, there will be apparent tensile stress concentration in the vicinity of dopant atoms, which will reduce the tensile failure strain and elastic modulus of graphene. The higher of the degree of aggregation of dopant elements, the greater of the reduction of tensile mechanical behavior of graphene, and the crack initiates from the location where dopant atoms aggregate. The defects introduced by doping have influence on the mechanical behavior of graphene structure, and the defects with high concentration can greatly reduce the tensile strength and fracture strain of graphene structure. Different types of elemental doping have different effects on the mechanical behavior of graphene.
4. The doped ions will inhibit the electronic transport properties of graphene. Different forms of doping have different effects on electronic transport properties and can produce transmission valleys at different positions of transport lines, which could lead to form donor and acceptor semiconductor. And the doping will destroy the symmetry of the graphene structure, resulting in an increase of the current through the graphene nanoribbon structure under bias. The location of the dopant elements can also affect the electronic transport properties.

References

1. Wu X, Zhao HY, Yan D, Pei JY (2017) Doping of graphene using ion beam irradiation and the atomic mechanism. Comp Mater Sci 129:184–193
2. Wei D, Liu Y, Wang Y et al (2009) Synthesis of N-doped graphene by chemical vapor deposition and its electrical properties. Nano Lett 9:1752–1758
3. Bangert U, Pierce W, Kepaptsoglou DM et al (2013) Ion implantation of graphene-toward IC compatible technology. Nano Lett 13:4902–4907
4. Wang H, Wang Q, Cheng Y et al (2012) Doping monolayer graphene with single atom substitutions. Nano Lett 12:141–144
5. Wang J, Liu Z (2012) First-principles study of the transport behavior of zigzag graphene nanoribbons tailored by strain. AIP Adv 2:012103
6. Li Z, Qian H, Wu J et al (2008) Role of symmetry in the transport properties of graphene nanoribbons under bias. Phys Rev Lett 100:206802
7. Biel B, Blasé X, Triozon F et al (2009) Anomalous doping effects on charge transport in graphene nanoribbons. Phys Rev Lett 102:096803

Chapter 5
Joining of Graphene by Particle Beam Irradiation and Its Properties

5.1 Introduction

Graphene applications are often related to graphene joining, and the molecular junctions between graphene and other nano-materials. On the one hand, the shape and area of the original graphene can be controlled by joining method. On the other hand, through joining the microelectronic components such as the graphene quantum dot, the PN junction or the Schottky junction can be assembled to prepare the three-dimensional complex structures. At present, there are few studies on the joining of graphene. While there are some specific studies on the joining of carbon nanomaterials, such as the connection between CNTs, the joining between graphene and CNT, the molecular junction between CNT and fullerene. The study of the joining process and mechanism of carbon nanomaterials by physical or chemical methods can provide a good inspiration and guidance for the work of this thesis. Particle beam has the advantages of small focusing radius, high processing precision, high processing efficiency, large energy regulation range and good controlled deflection flexibility, which have great potential application in the joining of graphene nanomaterials. Therefore, this work proposed the method of particle beam (laser and ion beam) irradiation method to join graphene, by which we hope to further promote the application of particle beam in graphene processing, and promote the development of graphene-based components.

5.2 Experimental Studies of Graphene Joining by Particle Beam Irradiation

In this section, low-moderate energy ions and laser were experimentally used to irradiate the overlapped graphene samples, and the joining possibility and joining mechanism were analyzed. The effect of the annealing process on graphene joining

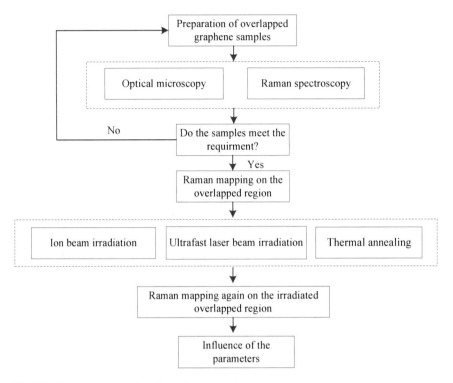

Fig. 5.1 Experiment procedure flow of graphene joining

was further studied by thermal annealing experiment. The experiment procedure of graphene joining is shown in Fig. 5.1.

5.2.1 Preparation and Characterization of Graphene Joining Samples

The preparation of the overlapped graphene test samples is as follows:

1. Preparation of single layer graphene by CVD method, see Sect. 2.1 for details.
2. Transferring the first monolayer graphene sample to a silicon substrate with 300 nm thickness oxide by etching, and then transferring the second monolayer graphene thereon. The two graphene sheets overlapped as shown in Fig. 5.2, for which the overlapped region has an area of 0.5×0.5 cm^2, which is 50% of the monolayer graphene.
3. Optical microscopy characterization of overlapped graphene samples. Figure 5.3 shows the image of the overlapped graphene under optical microscope. It can be seen that there is a clear dividing line between the overlapping

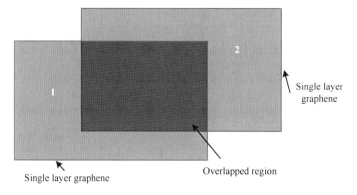

Fig. 5.2 Illustration of the overlapped graphene sample

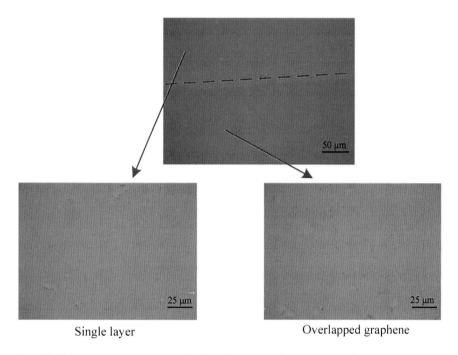

Fig. 5.3 Optical microscopy characterization of overlapped graphene sample

area and the non-overlapping area (where is marked by the dotted line). When observed at high magnification, the overlapping region and the non-overlapping region did not show much difference.

4. Raman spectroscopy characterization of overlapped graphene samples. Figure 5.4 shows the Raman spectroscopy characterization results of the graphene with overlapping region and the non-overlapping region. From Fig. 5.4a,

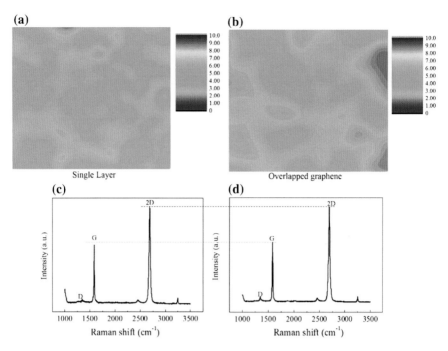

Fig. 5.4 Raman characterization of graphene samples in non-overlapping region and overlapping region. **a** The distribution of Raman peak intensity ratio (I_{2D}/I_G) of the non-overlapping region. **b** The distribution of Raman peak intensity ratio (I_{2D}/I_G) of the overlapping region. **c** Typical Raman spectrum signal in non-overlapping region. **d** Typical Raman spectral signal in overlapping region. Raman imaging area is 50×50 μm^2, with 51×51 imaging points, i.e., sampling at about every 1 μm. Reprinted with permission from Ref. [4]. Copyright 2017, AIP Publishing LLC

b, it can be seen that the distribution of Raman peak intensity ratio (I_{2D}/I_G) in the overlapped and non-overlapping regions is substantially kept at a nearly uniform value (~ 3 to 5). The Raman signal of graphene samples can characterize the difference between monolayer graphene and multilayered graphene. The Raman peak intensity ratio (I_{2D}/I_G) of our monolayer graphene sample is between 3 and 5, which is consistent to that of Raman signals of normal monolayer graphene, and our sample exhibits a uniform result in large area. The Raman peak intensity ratio (I_{2D}/I_G) of the overlapped sample is still in the range of about 3–5, which presents a signal of monolayer graphene. There are two possible causes for this phenomenon [1–3]: On the one hand, the crystal rotation angle between the two overlapped graphene sheets is too large or too small. On the other hand, the separation distance between the two layers is too big. In this paper, the overlapped sample is prepared by laying the monolayer graphene one by one, which would result in a large spatial distance between the two layers. Meanwhile, during the transfer process, there will be some residual impurity elements and graphene fragments easily existing in the interlayer, which could also lead to a large separation distance between the two layers. The large interlayer spacing is the main reason for the existence of consistent Raman signals in the overlapping region and non-overlapping region. So that the coupling between the two layers is too weak, which makes it difficult to transfer the carrier. Anyway, the consistent Raman signal between the non-overlapping region and the overlapping region indicates that there is almost no coupling (joining) between the two graphene formed before processing. In this paper, we also did the typical single-point Raman spectra of the non-overlapping region and the overlapping region, as shown in Fig. 5.4c, d. It can be seen from the figure that Raman signals for both the overlapping region and the non-overlapping region show consistent single layer information.

5. AFM characterization of overlapped graphene samples. The results show that (as shown in Fig. 5.5): (a) The whole part of the prepared graphene sample is relatively uniform, and there is a clear dividing line between the overlapping region and the non-overlapping region; (b) The interval spacing between different layers is around 1–2 nm for our specimen, which is much larger than the interval spacing of the as-grown double layer graphene (~ 0.35 nm); (c) The element composition for the overlapping region and the non-overlapping region is uniform (both are only carbon element). The experimental results of AFM

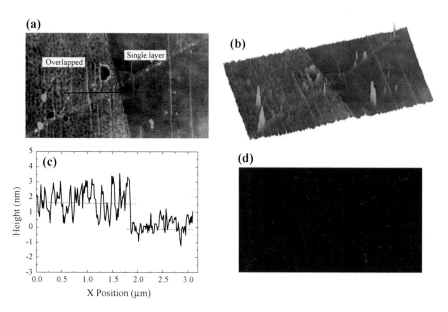

Fig. 5.5 AFM characterization of overlapped graphene sample. **a** The height profile of the overlapping area and the non-overlapping region. **b** The corresponding three-dimensional topography. **c** The height distribution along path L as shown in (**a**). The red dashed lines represent the average height. **d** The phase diagram containing the overlapping region and the non-overlapping region. The scanned area is 30 μm. Reprinted with permission from Ref. [4]. Copyright 2017, AIP Publishing LLC

further show that there is a large interlayer distance for the overlapping region, so the interaction force between the layers is weak.
6. Repeat steps 1–5 until a sample satisfying the experimental requirements is obtained: there is a large area of the overlapping region; and the Raman signal of the overlapping region is substantially consistent with the non-overlapping region; the atomic force signal shows there is a large spacing distance between the overlapping region and the non-overlapping region.

5.2.2 Graphene Joining by Ion Beam Irradiation

After the preparation of graphene samples, the possibility of joining of the overlapped graphene samples was studied. In this section, low-energy nitrogen ion beam was used to irradiate the overlapped graphene samples, and the Raman spectra of the overlap region before and after irradiation were observed. The experimental parameters of low energy ion implantation were as follows: the gas flow rate was 250 mL/min, and the upper power was 800 W. The ion implantation time was 10 s, with a pulse width as 30 μs, and the injected ion energy was 20 and 40 eV, respectively. Under these parameters, the dose of the injected ion beam is about $1 \times 10^{-14}/cm^2$.

Figure 5.6 shows the Raman signal in the graphene overlapping region before and after ion beam irradiation. The distribution of Raman peak intensity (I_{2D}/I_G) of the overlapped graphene samples before ion beam irradiation uniformly ranges from 3 to 5, which indicates that the samples are homogeneous and the interlayer force is weak. After ion beam irradiation, the Raman peak intensity ratio (I_{2D}/I_G) of the sample ranges from 0 to 2 in large area, which indicates that the interlayer force of graphene after ion beam irradiation has changed greatly, and the original overlap region begins to show the signal of multi-layer graphene. There is enhanced coupling contact (joining) occurred in some areas between the two graphene sheets in overlapped region. Figure 5.6c further shows that the Raman peak intensity ratio (I_{2D}/I_G) decreases significantly after irradiation in most area.

Compared with the literature [1–3], it can be seen that the Raman signal of the original graphene in overlapped region shows that the contact between the graphene layers is weak. The Raman signal of the overlap region after ion beam irradiation shows that the interaction is obviously enhanced, indicating that ion beam irradiation changes the interaction between graphene layers, which may produce the effect of "joining".

After the ion beam implantation, the ions collide with the graphene lattice, and the carbon atoms in the upper graphene are affected by the downward impact force, which leads to the decrease of the graphene interlayer spacing and the torsion of the graphene structure, so that the crystal orientation deflection angle of the graphene interlayer changes, thereby increasing the coupling contact between the graphene layers. Figure 5.7 shows the typical single-point Raman spectra of the overlap

5.2 Experimental Studies of Graphene Joining by Particle Beam Irradiation

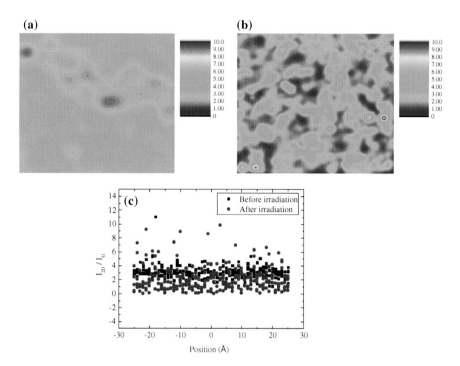

Fig. 5.6 Raman mapping spectra of the overlapped region before and after 20 eV of nitrogen ion irradiation. **a** The distribution of the Raman peak intensity ratio (I_{2D}/I_G) before irradiation. **b** The distribution of the Raman peak intensity ratio (I_{2D}/I_G) after irradiation. **c** The corresponding values of I_{2D}/I_G. Only 500 original data points were taken in (**c**). Reprinted with permission from Ref. [4]. Copyright 2017, AIP Publishing LLC

region before and after ion beam irradiation. It can be directly seen that the Raman spectrum peak intensity ratio (I_{2D}/I_G) is significantly reduced after ion beam irradiation. Meanwhile, there are obvious defect peak (D peak) generated after ion irradiation, which indicates that the low energy ion beam implantation will bring defects to the graphene structure together with the joining of graphene. According to the results of the graphene doping by low-energy ion implantation, it is known that the irradiation of low energy ions to the overlapped graphene samples will lead to substitution, adsorption of the graphene samples, and also bring vacancy defects to the overlap region. Besides, some of the incident atoms are embedded in the interlayer after low energy ion implantation. As for the microscopic phenomena and atomic mechanisms of graphene joining by low-energy ion implantation, MD simulations will be used to discuss it in the following section. When overlapped graphene sample is irradiated by a higher energy (40 eV) ion beam, the Raman peak ratio (I_{2D}/I_G) of the graphene sample will still be significantly reduced, indicating that when increasing the energy of ion beam, we can still get the joining results. However, under the action of higher energy ion beam, the defective peak in graphene is more obvious, which shows that increasing the energy of the ion beam

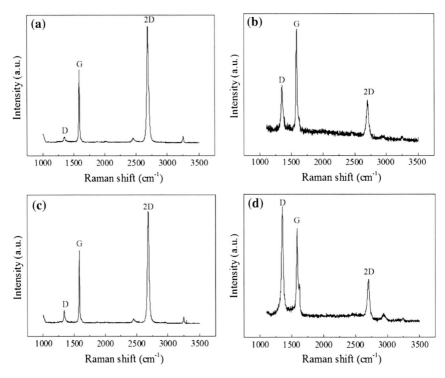

Fig. 5.7 Typical single point Raman spectra of the overlap region. **a** and **c** Before ion beam irradiation. **b** and **d** After ion beam irradiation. The ion beam energy is 20 eV (**b**), 40 eV (**d**), respectively

will bring more defects to the graphene structure. The effect of ion beam irradiation parameters on graphene joining results will be discussed in subsequent MD simulations.

5.2.3 Graphene Joining by Laser Irradiation and Thermal Annealing

Because the ion beam irradiation is conducted at room temperature and pressure, there is no significant thermal effect for graphene samples. The joining effect of graphene by ion beam irradiation is generated by the collision between the incident ion beam and graphene carbon atoms. In order to further explore the possibility of graphene joining under thermal action, laser and thermal heating were used in this section to try to join the overlapped graphene, and the joining signal was characterized.

For the laser induced joining of graphene, the results were taken from Ref. [1].

5.2 Experimental Studies of Graphene Joining by Particle Beam Irradiation

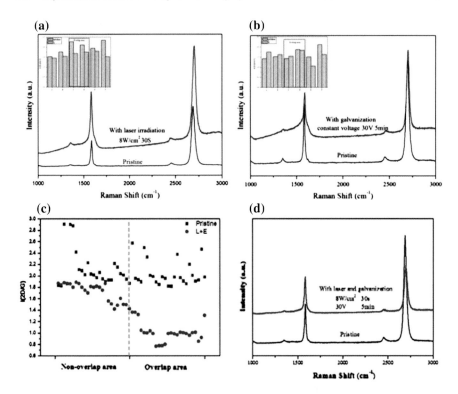

Fig. 5.8 Raman spectra of overlapped region under the action of laser and current. **a** Raman spectra before and after laser action. **b** Raman spectra before and after current action. **c** Raman peak intensity ratio (I_{2D}/I_G) of the overlapped region under the combined action of laser and current. **d** Typical single point Raman spectra under the combined action of laser and current. Reprinted from Ref. [1], Copyright 2013, with permission from Elsevier

Figure 5.8 shows the Raman signal changes of overlapped graphene under the action of continuous CO_2 laser and current Joule heat. From Fig. 5.8a, b, it can be seen that the coupling effect between the overlapped graphene sheets is still weak under the action of laser alone and current Joule heat alone, which means the laser or current cannot generate the joining results as the ion beam irradiation. While in the case of combined continuous laser and current, the Raman signal peak intensity ratio (I_{2D}/I_G) of graphene changes significantly before and after action. In fact, the I_{2D}/I_G value decreased after the action, which indicated the coupling effect between graphene is obviously strengthened, and the molecular junctions between graphene layers are formed. Reference [1] explained the joining signal of overlapped graphene under the combined action of laser and current as follows: under the combined action of laser and current, there will be O_2 and H_2O molecules adsorbed in the interlayer, which could make the two layers be closer, and reduce the repulsive interaction between layers, resulting in an enhancement effect of the π-π coupling. This study demonstrated that the combination of laser and current action can also result in the joining phenomenon of overlapped graphene.

Since graphene only absorbs a small portion of the laser energy (2.3% per layer), the thermal phenomenon of graphene under laser action is relatively weak. In order to further observe the effect of the thermal process on the joining of overlapped graphene, this work then studied the joining possibility of graphene by thermal annealing it in a vacuum chamber. Figure 5.9 shows the thermal history of the graphene samples. For which, the insulation temperature is 300 °C, and the holding time is about 2 h, with 1×10^{-3} Pa vacuum condition.

Figure 5.10 shows the typical Raman spectra of the non-overlapping region and the overlapping region before and after vacuum thermal annealing. The graphene Raman signal does not change in the non-overlapping region after thermal annealing, which shows the thermal annealing process has little effect on the atomic structure of the non-overlapping region. While the graphene Raman signal in the overlapping region has changed greatly: the G peak intensity is obviously enhanced, and Raman intensity ratio (I_{2D}/I_G) is significantly reduced. These indicate that the coupling effect between the graphene layers after vacuum thermal annealing has been enhanced, so vacuum annealing can also produce the "joining" effect of the overlapped graphene.

Figure 5.11 shows the distribution of the Raman signal peak intensity ratio (I_{2D}/I_G) before and after the vacuum thermal annealing. It can be seen from the figure that the overlapped sample show a relatively uniform Raman signal of monolayer graphene before heat treatment, indicating that the coupling effect between the graphene layers is weak. After the vacuum heat treatment, the Raman peak ratio (I_{2D}/I_G) showed a reduction phenomenon in large area, indicating that the vacuum thermal annealing process can increase the coupling between the graphene layers. Moreover, compared to the ion beam irradiation, the distribution of I_{2D}/I_G under thermal annealing is more uniform and the additional defective peak (D peak) is not increased.

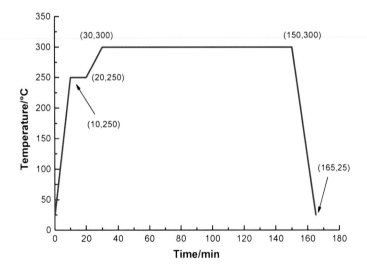

Fig. 5.9 Temperature history during the thermal annealing of overlapped graphene samples

5.2 Experimental Studies of Graphene Joining by Particle Beam Irradiation 109

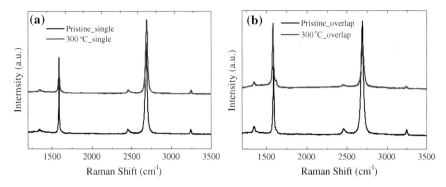

Fig. 5.10 Comparison of typical Raman signals before and after vacuum heat treatment for samples in **a** non-overlapping region and **b** overlapping region

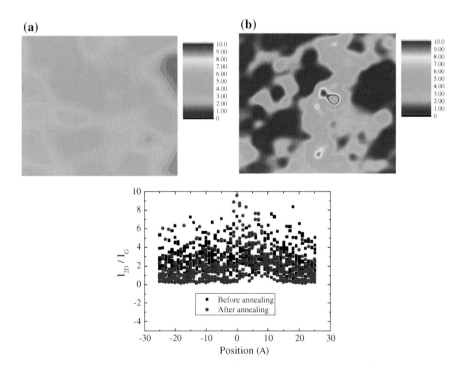

Fig. 5.11 Variation of Raman spectroscopy signals for overlapped region before and after heat treatment. The distribution of the Raman peak intensity ratio (I_{2D}/I_G) **a** before and **b** after heat treatment. **c** The corresponding values of I_{2D}/I_G. In order to show clearly, only 1250 original data points were used in (**c**)

Unlike the ion beam irradiation and combined laser-current action, vacuum thermal annealing of graphene joining is mainly due to the thermal effect. The original graphene samples exhibit a large interlayer spacing due to the presence of some impurity elements and graphene fragments between the layers. Under the action of heat treatment, these impurity atoms and the embedded carbon atoms would migrate [5, 6] and may eventually escape from the interlayer region, which could reduce the contact distance between the graphene layers and increase the coupling contact. At the same time, graphene crystal structure will expand and twist under heat treatment, and the difference of the thermal expansion coefficient between SiO_2 substrate and graphene will also lead to a significant torsion of the graphene structure, so the difference of the crystal orientation angles between the two layers may come to a value to induce the obvious increase of the G peak as stated in literatures [2, 3], which could also lead to the decrease of the I_{2D}/I_G value. As the laser absorption efficiency of graphene is very low, there is no obvious thermal effect for graphene samples under laser action, so the laser action alone is difficult to produce such element migration and structural torsion.

5.2.4 Experiment Summary

By using the methods of ion beam irradiation, laser irradiation and thermal annealing to study the graphene joining process, the following conclusions can be drawn:

1. Ion beam irradiation can be used to join overlapped graphene; under the action of laser irradiation alone, it is difficult to form the molecular junctions between graphene sheets. While with the auxiliary role of current, the laser can be used to join overlapped graphene; thermal annealing process can also induce the joining results of the overlapped graphene.
2. The joining of graphene by ion beam irradiation is caused by the collision between the incident ions and graphene carbon atoms; the joining of graphene by combined laser-current action is due to the enhanced π-π coupling effect between graphene layers induced by O_2 and H_2O adsorption; thermal annealing induced graphene joining is formed by the element migration and structural torsion resulted from annealing process. In-depth joining phenomena and mechanisms need to be further clarified.
3. Ion beam parameters and laser power and other factors have an impact on the joining results.

5.3 Theoretical Analysis of the Joining Mechanism

Both the ion beam irradiation and the laser action can induce the phenomenon of enhanced molecular coupling between graphene layers. Due to the particle interaction effect of the ion beam and the optical fluctuation characteristics of the laser photon, the graphene joining mechanism will not be the same under different

5.3 Theoretical Analysis of the Joining Mechanism

particle beams irradiation. In this section, MD simulation was used to study the joining of graphene with different relative positions under the effect of ultrafast laser and ion beam irradiation, and explain the microscopic mechanism of graphene joining under different conditions. Due to the limitation of the simulation space scale (within the nanometer) and the time scale (within nanosecond), the MD simulation cannot reproduce the experimental conditions, and also the special method is needed to define the joining of graphene layers. Therefore, the simulation conditions were simplified as follows:

1. The size of the graphene sheet to be joined is on the order of nanometer.
2. The time period for laser and ion beam action is in the order of picosecond.
3. The laser is uniformly applied to the whole graphene region, and the ion beam irradiation acts on the graphene overlapping region.
4. For different overlapped graphene samples, it is assumed that the interstice between two graphene layers have reached the spacing of bilayer graphene (~ 0.34 nm), and the research focused on the effect of laser or thermal annealing, ion beam collision on the formation of chemical bonds between graphene layers.
5. The success or not of graphene joining is judged as whether or not a chemical bond can be formed between graphene sheets.

Since this work focused on the different mechanisms of graphene joining due to the different joining methods, and there are already a lot of work using similar hypothesis methods to study the joining of carbon nanomaterials [7–9], these assumptions are reasonable and can meet the needs. Meanwhile, it can greatly reduce the space and time scales of simulation.

5.3.1 Graphene Joining Under Laser Beam Irradiation

5.3.1.1 Research Model

In this section, we first studied the possibility of the formation of chemical bonds for graphene with different relative positions under the action of laser irradiation. The joining type includes overlap joining, butt joining, corner joining, dislocated butt joining and T-joint joining, in which the overlap and butt joining cases are mostly used in practical application, which are also the focus of this study. Figure 5.12 illustrates the models of five types of graphene joining under ultrafast laser irradiation. The size of the two graphene was 50×100 Å. Because the laser spot is much larger than the size of graphene sheet, the laser action was assumed to be a uniform heat source on the whole graphene body. Under this assumption, the model actually simulated the joining results of graphene under the thermal effect of the laser and annealing process. The time of laser action is 5 ps and the power is 4.29×10^8 W cm^{-2}. The ratio of laser energy absorbed by graphene is 2.3%.

Under this energy laser irradiation, the atomic structure of graphene is not destroyed. These conditions can be used to simulate the formation possibility of chemical bonds between graphene under 5 ps single pulse laser. The initial spacing between the different graphene sheets in the horizontal and vertical directions is 3.4 Å, and the armchair-type boundary of graphene was fixed. The establishment of the joining model was done in the VMD. First, the model of the monolayer graphene was established, and then the spacing and the angle between the two graphene were controlled by translation and rotation to obtain the desired relative positions. The graphene was equilibrated at room temperature (300 K) for a long time; then the laser heat source was applied and the structures of the two graphene changed under the action of laser. After the laser action, the system was freely relaxed in the NVE until the temperature of the graphene structure gets stabilized. Figure 5.12 also describes how to obtain a research sample of graphene joining under different relative positions in practical experiment. First, the monolayer graphene is placed between the two metal electrodes, and the metal electrode is electrically connected to the conductive circuit. The movement of the electrode can

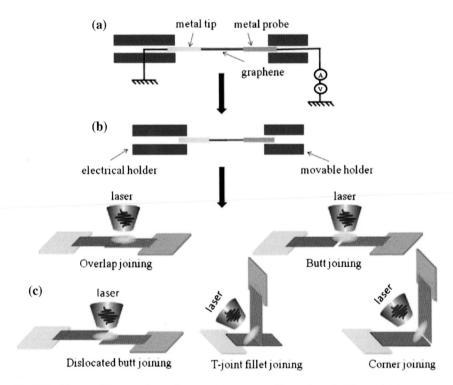

Fig. 5.12 The possible experimental setup and geometry illustration of different joining cases. **a** The single layer graphene is adhered to the metal base. **b** The single graphene sheet is broken into two segments using some methods. **c** Five different cases researched in the following studies. Reprinted with the permission from Ref. [11]. Copyright 2013 The Japan Institute of Metals and Materials

5.3 Theoretical Analysis of the Joining Mechanism

be precisely controlled by the micro-manipulator, and the current flowing through the graphene can be used to promote the connection between graphene and metal electrodes; and then the FIB or focused electron beam, or increasing the external voltage, or other methods is used to "cut off" the original graphene between the metal electrodes, making it two separate graphene; and then the relative positions between the two pieces of graphene are further controlled through the movement of one end of the micro-mechanical arm. From these steps, the experiment specimens for overlap joining, butt joining, corner joining, dislocated butt joining, and T-joint joining cases can be obtained. For more details about the experimental procedure, please refer to Ref. [10].

5.3.1.2 Results of Graphene Joining Under Laser Irradiation

Figure 5.13 shows the simulated results for graphene joining under ultrafast laser. It can be seen from the figure that for the types of overlap, T-joint, two graphene sheets are still independent of each other, it cannot form an effective chemical bond between them, so there are no new chemical bonds formed under laser action, the information of layer number increase for overlapped graphene from experiment is only the result of enhanced intermolecular force. While for the cases of corner, butt and dislocated butt relative positions, there are new chemical bonds generated. However, for the corner type, there are many polygon defects appeared in the joint. While for the butt and dislocated butt types, the joint quality is very good. In fact, it is also possible to observe the effect of the distance between the two pieces of graphene in the horizontal and vertical directions on the joining results. When the distance between the two graphene is too large, all joining attempts are not successful. With the reduction of the distance (the critical distance is related to the laser power and the size of the graphene sheet, which will be discussed detailedly in future studies), attempts for corner, butt and dislocated butt types can be successful, while the overlap and T-joint types joining trials are always unsuccessful. Table 5.1 summarizes the results of the above joining trials. The different joining results under different relative positions will inevitably lead to the discussion of the mechanism of the formation of atomic bonding between graphene under laser irradiation.

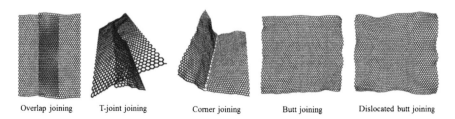

Overlap joining   T-joint joining   Corner joining   Butt joining   Dislocated butt joining

Fig. 5.13 Simulation results of graphene joining under laser irradiation

Table 5.1 Possibilities of the formations of molecular junctions for different joining cases

Joining cases	Molecular junctions (Yes/No)
Overlap joining case	No
Butt joining case	Yes
Dislocated butt joining case	Yes
T-joint fillet joining case	No
Corner joining case	Yes

Reprinted with the permission from Ref. [11]. Copyright 2013 The Japan Institute of Metals and Materials

5.3.1.3 Mechanism of Graphene Joining Under Laser Irradiation

Unlike most solid materials, graphene has a negative thermal expansion coefficient at room temperature along the direction parallel to the sp^2 hybridization [12], and under freely suspended condition, graphene is confirmed to have a lattice constant reduced with the increase of temperature if the temperature ranges from 500 to 700 K [13]. Thus, if one would like to obtain the results of the molecular junctions between graphene by means of thermal expansion, one of the ways is to continuously increase the energy of the laser to cause the graphene sheets to expand, thereby "crossing" the initial spacing to form a joining. While too large laser energy is likely to cause the destruction of the atomic structure of graphene. It can be seen from the study in Chap. 3 that the expansion and contraction of the graphene in the plane and fluctuation out of the plane will increase under the action of laser heat, which are also reported in the literature [14]. Figure 3.4 shows the length of graphene in the plane and displacement of the edge atom in out of plane direction under different temperature field. It can clearly see the expansion and contraction and fluctuation of graphene structure under certain temperature, and these phenomena will be enhanced with the increase of temperature. It can be presumed that when the momentary elongation of the graphene sheet in the plane exceeds the initial spacing, it is possible to form chemical bonds between the graphene in the plane direction. And also when the fluctuation of graphene sheet in the direction perpendicular to the plane exceeds the initial spacing, it is possible to form chemical bonds between the graphene in the out of plane direction. Of course, the formation of chemical bonding must also overcome a certain energy barrier, which makes it be hard for the graphene to automatically bond at room temperature.

Thus, the expansion and contraction of the graphene structure in the plane and the fluctuation out of the plane provide the "driving force" for the formation of chemical bonds between two graphene with different relative positions, and also enable them to overcome the initial energy barrier. However, under the action of this "driving force", overlap joining and T-joint joining cannot be successful, which indicated that besides this "driving force", there must be some other decisive factors.

In order to explain the graphene joining mechanism, this work further analyzed the two most common types of joining—overlap and butt joining, and other types have similar mechanisms. Figure 5.14 shows the comparison between the

5.3 Theoretical Analysis of the Joining Mechanism

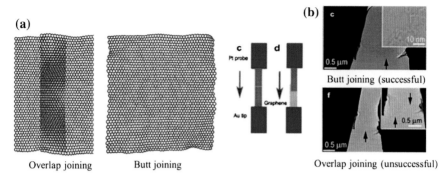

Fig. 5.14 Comparison of graphene joining results between **a** simulation and **b** experiment. Reprinted from Ref. [10], Copyright 2012, with permission from Elsevier

simulation results and the experimental results for the overlap joining and butt joining cases, where the experimental results are from Ref. [10]. It can be seen from the figure that the simulation is in agreement with the experimental results, that is, there is no chemical bonding formed under the overlap joining type, while there is chemical bonding formed between the graphene sheets under the butt joining type. It can be seen from Fig. 5.15 that the carbon atoms at the edge of graphene have dangling bonds and the internal carbon atoms are saturated. So that the carbon atoms with the dangling bonds tend to be saturated and the internal carbon atoms do not have this "drive force". However, in order to generate the joining results, both of the two pieces of graphene must have dangling bonds. For the case of butt joining, each of the graphene has the carbon atoms at the edge with dangling bonds, which could lead to the formation of chemical bonding by the saturation of the dangling bonds. While for the case of overlap joining, due to the two graphene interact in the internal region, which means it cannot provide carbon atoms with dangling bonds for the upper and lower graphene at the same time, so there is no chemical bonding for the overlap joining under ultrafast laser and heating action.

Therefore, under ultrafast laser action, the expansion and contraction of graphene in the plane and the fluctuation out of the plane provide the "driving force" for the joining of graphene, and the final joining results are determined by the

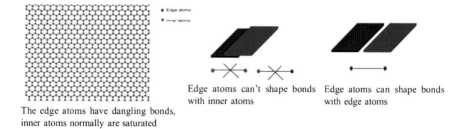

Fig. 5.15 Mechanism of graphene joining under ultrafast laser action

saturation effect of the carbon atoms with dangling bonds at the edge of graphene. Under this mechanism, the overlap and T-joint types cannot generate the chemical bonding between graphene layers, while butt, dislocated butt and corner types can lead to the chemical bonding. The joining signal for overlapped graphene under laser action obtained from experiment is actually the effect of enhanced intermolecular force, other than the formation of chemical bonds.

5.3.2 Joining of Graphene by Ion Beam Irradiation

5.3.2.1 Research Model

Compared to the laser, ion beam irradiation will generate a collision between atoms, and lead to structural damage, transformation, reorganization, which may bring different joining results to graphene. In order to study the possibility and influence factors of chemical bonding between graphene sheets under the action of ion beam, the joining of overlapped graphene was studied by using the model shown in Fig. 5.16, and the change of graphene structure under ion beam irradiation was also discussed.

The size of two graphene in Fig. 5.16 is 50 × 100 Å, and 50 Å area is overlapped along the length direction, which means half of the whole size is overlapped. Fixed boundary conditions was applied for the armchair-type edges, and five layers of carbon atoms were selected as part of the thermal bath area near the edge to absorb the stress wave formed by the ion beam collision. Due to the limitation of space and time scale, the area irradiated by the ion beam was confined to the overlap region to simulate the joining of the overlapped graphene under the action of FIB. The energy of the ion beam was controlled by the kinetic energy of the incident ions, and the dose of the ion beam was determined by the number of

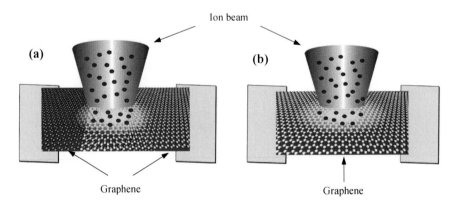

Fig. 5.16 Study models of **a** overlapped graphene and **b** monolayer graphene irradiated by ion beam. Reprinted from Ref. [15], Copyright 2014, with permission from Elsevier

incident ions per unit area. The influence of different parameters on the joining results was discussed here, where the energy distribution of the ion beam was 20–100 eV and the dose distribution of the ion beam was 2.65×10^{14}–2.12×10^{15} ions/cm^2. The incident ions were initially located in a cubic box 40 Å above the graphene plane, and then an ion was incident every 500 MD time step until the desired dose was reached. After irradiation, high temperature annealing process was conducted for the system. The annealing temperature is 2000 K and the time period is 20 ps. The annealing temperature and time can represent the low temperature annealing in the actual process for a long time. Finally, the system was cooled to room temperature and then the results were extracted. AIREBO potential was selected to describe the interaction between the carbon atoms in graphene. And the interaction between the incident carbon ions and the carbon atoms in graphene was described by Tersoff potential, which is smoothly connected with the ZBL exclusive potential in short distance to represent the collision between different atoms. For the collision between Ar ions and carbon atoms in graphene, it was described only by ZBL exclusion potential because there is no chemical bonding between them. The incident ions selected were C ion and Ar ion, respectively. The C ion can represent the type of ions that can be chemically bonded to the graphene. Ar ion can represent the ionic species that cannot be chemically bonded to the carbon atoms in graphene. For the incident nitrogen ion used in the experiment, it belongs to the type of ions that can form chemical bonding with graphene, so graphene joining by nitrogen ion irradiation has the same mechanism as that of carbon ion. The charge of the ions was not taken into account for all the simulations.

5.3.2.2 Joining Results of Graphene by Ion Beam Irradiation

Figure 5.17 shows the changes in the structure of the overlapped graphene after FIB irradiation, which is also compared to the results of the laser action. Under the action of laser beam, though the graphene structure will crimp and fold, the graphene hexagonal honeycomb atomic structure is still maintained. While under the action of FIB, the atomic structure of the graphene in overlapped region changes. Part of the original hexagonal honeycomb structure transforms to be an amorphous structure. In order to evaluate the graphene joining results under laser and ion beam irradiation, this section also did the uniaxial tensile test simulation for the

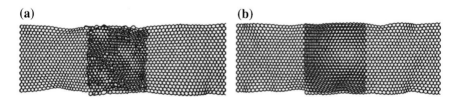

Fig. 5.17 Changes of the structure of overlapped graphene under the action of **a** FIB irradiation and **b** laser heat. Reprinted from Ref. [15], Copyright 2014, with permission from Elsevier

as-joined graphene. For which the tensile process was carried out at room temperature (300 K), and the simulation software was LAMMPS. During the stretching process, the graphene was fixed at one end and the other end moved at a strain rate of 0.0002 ps^{-1} in a direction parallel to the zigzag boundary. See Chap. 2 for the extraction method of atomic stress after stretching.

Figure 5.18 is the tensile simulation results of overlapped graphene under laser and ion beam action. From the stress-strain curve, it can be seen that under laser action, the interaction force between the graphene sheets is very weak and the stress is always close to zero. At this time, the interaction force between the graphene sheets is van der Waals force. The study of CNT joining [8] also pointed out that high temperature can promote the transformation of nanotube structure and defect annealing, but cannot form a connection between the nanotubes. While under the irradiation of FIB, the tensile stress of overlapped graphene first increases as the strain increases, and eventually reaches its maximum, and then the stress decreases sharply as the strain increases. This stress-strain relationship coincides with the normal tensile-elongation-failure process as the strain changes, confirming that there are strong bonds generated between the two sheets under the ion beam irradiation. This interlayer force is much larger than the intermolecular force existing in the pristine overlapped sample.

5.3.2.3 Joining Mechanism of Graphene Under Ion Beam Irradiation

In order to explain the reason of atomic bonding in graphene under the irradiation of ion beam, it is necessary to clarify the change of graphene structure under ion beam. Figure 3.9 shows the change of graphene structure under ion beam irradiation. Since the damage threshold of the graphene structure under ion beam irradiation is about 20 eV [16–18], most of the incident ions are present in the graphene structure by adsorption or substitution due to the weak destruction of the graphene structure at 35 eV. When the energy of the ion beam is large (200 eV), the incident

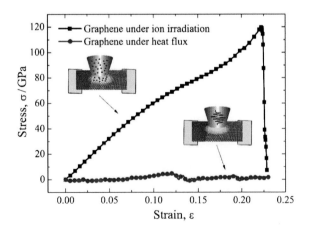

Fig. 5.18 Tensile simulation results of overlapped graphene under laser and ion beam irradiation. Reprinted from Ref. [15], Copyright 2014, with permission from Elsevier

5.3 Theoretical Analysis of the Joining Mechanism

ions will knock out the carbon atoms in graphene, resulting in a large number of defects in the graphene structure. These defects are mainly vacancies, which could change the original sp^2 saturated carbon atoms in the graphene structure into unsaturated, and these unsaturated carbon atoms tend to be saturated. Meanwhile, the adsorption and embedding of the incident ions will change the hybrid form of some carbon atoms in original graphene, resulting in the formation of sp^3 hybrid. Therefore, the types of the interaction force between the original overlapped graphene layers may be changed.

For the overlapped graphene, this section gives the morphology of the graphene after ion beam irradiation using ball and stick model, as shown in Fig. 5.19. It can be seen from the figure that after irradiation, there are a large number of incident ions embedded in the graphene overlap region, and the hexagonal shape honeycomb structure of the original graphene in the overlap region changes to be amorphous. The cross-sectional view in Fig. 5.19a shows that there are obvious chemical bonds formed between the carbon atoms in the upper and lower layers in the overlap region. When taking a careful look at the formed chemical bonds, it is found that there are two categories for the chemical bonds formed between the graphene layers: one is constituted by the carbon atoms in the graphene. The carbon atoms in the upper graphene move downward after the impact, and overcome the initial energy barrier between the graphene layers to generate chemical bonds. During the collision process, the hybrid form of the carbon atoms in graphene will be changed from saturated sp^2 to unsaturated sp or sp^3. As we know that these unsaturated carbon atoms tend to be saturated, and easily generate new chemical bonding with the lower graphene. Therefore, this kind of chemical bonding between the carbon atoms in graphene is called "coordination joining". The other one is that the incident ions are embedded between the graphene layers to form a kind of "bridge" effect. That is, the embedded ions change the hybridization of

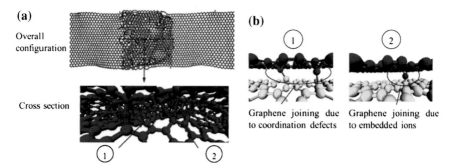

Fig. 5.19 Joining mechanism of graphene under ion beam irradiation. **a** The ball-stick model representing the joining results. **b** Two joining mechanisms. Wherein the blue and red balls represent the carbon atoms in graphene and the incident ions, respectively. The C–C bonds in graphene are represented by blue sticks. The chemical bonds between the incident ions and the graphene are indicated by the red sticks. The length of the carbon-carbon bonds is 2.0 Å. Reprinted from Ref. [15], Copyright 2014, with permission from Elsevier

some carbon atoms in the original graphene and then bond with the carbon atoms in the upper and lower graphene simultaneously to form new chemical bonds. Thereby the two graphene sheets are linked by these bonds. This type of graphene joining phenomenon due to the bridging of embedded ions is called as the "embedded ions joining".

Thus, the joining of overlapped graphene under the irradiation of the ion beam is reflected by the generation of new chemical bonds, which is mainly caused by two mechanisms, i.e. "coordination joining" and "embedded ions joining". Figure 5.19b is the illustration of two different joining mechanisms. Observing the chemical bonds formed between the graphene layers, it is found that the bridging effect of the embedded ions always plays a leading role in the process of graphene bonding in the studied energy range (20–100 eV). It can be summarized that for the experimental ion species that can form the chemical bonds with carbon atoms in graphene, such as carbon ion, nitrogen ion, boron ion, silicon ion, etc., there are always two joining mechanisms existed for overlapped graphene under the incident of ions. While for those ion species that cannot chemically bond with carbon atoms in graphene, such as Ne, Ar, Ke, Xe ion, etc., they will not form a bridging effect between the graphene layers after incident, so only the "coordination joining" mechanism exists.

5.4 Mechanical Properties of the Graphene Joint

In the first two sections, the experimental and theoretical results of the joining of graphene under laser and ion beam irradiation were discussed. In this section, the mechanical properties of graphene joint were studied to evaluate the application of the obtained graphene joint in the mechanical structures.

5.4.1 Mechanical Properties of Butt Joint of Graphene

5.4.1.1 Research Model

According to the results of Sect. 5.3, under the action of laser, the graphene sheets with relative positions of overlap and T-joint cannot form atomic bonding between each other, while the graphene sheets with butt, dislocated butt and corner types can form atomic bonding. The interaction force between graphene sheets is very weak if no chemical bonds are formed, for which it is meaningless to research its mechanical properties. For the cases that atomic bonds can be generated, it is actually impossible to obtain the ideal joint in actual process. This section analyzes the most common butt type in the actual process. The MD model is shown in Fig. 5.20, where the effects of chirality and crystal orientation of the two graphene sheets were discussed.

5.4 Mechanical Properties of the Graphene Joint 121

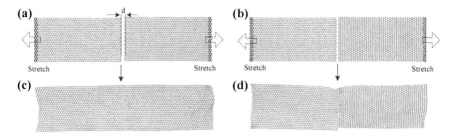

Fig. 5.20 Research models of the mechanical properties of butt joint. **a** Butt joining between zigzag graphene. **b** Butt joining between armchair and zigzag graphene. **c** and **d** are the corresponding joint after laser action. d is the distance between the two pieces of graphene before joining. The chirality is defined as the boundary along the tensile direction

In Fig. 5.20, the size of two graphene is 50 × 100 Å, and butt joint with different forms is generated between two graphene under laser action. The joint generated between two pieces of zigzag graphene is called Z-Z joint, and the joint generated between armchair type and zigzag type graphene is called Z-A joint. Z-Z joint can form a hexagonal structure consistent to original graphene. While due to the mismatch of the boundary atoms in Z-A joint, there are polygonal defects generated in the Z-A joint. Uniaxial tensile test simulation was conducted for the investigation of mechanical properties, in which the carbon atoms in the graphene edge layer moved along the horizontal direction at a strain rate of 0.001 ps^{-1}. The interaction between carbon atoms was described by AIREBO. The initial distance between graphene will change the interior strain presenting in the joint, which would affect the mechanical properties of the joint. Here, the initial spacing d was set to 3.0 Å, and the effect of d on mechanical properties will be specifically discussed in future study.

5.4.1.2 Mechanical Properties of Butt Joint with Different Forms

Figure 5.21 shows the tensile mechanical properties of the two graphene when they are joined at different chiral interfaces. It can be seen from Fig. 5.21a that the mechanical properties of the graphene joint are basically the same as that of the original graphene when the two graphene are joined at the same chiral interfaces (Z-Z). During the stretching process, there is no locally weak region of the whole joined graphene, so that the failure location of the graphene is random. The results show that the tensile damage originated at the edge of the graphene when a perfect joint is formed between the two pieces of zigzag graphene, indicating that it is possible to form a perfect joint for two pieces of butted graphene in the ideal situation. When the two graphene are joined at different chiral interfaces, or there are crystal orientations between two graphene, there will be vacancy defects generated due to atomic mismatch at the butt joint position, which will be the weak position of stress concentration under uniaxial stretching. Figure 5.21b is the tensile

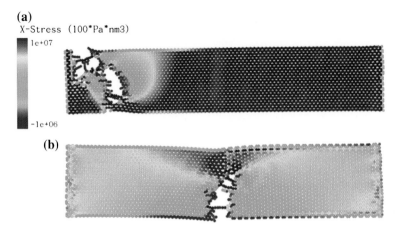

Fig. 5.21 Stress distribution of joined graphene before the tensile fracture in the case of **a** the Z-Z joint and **b** Z-A joint

test results of two butted graphene joined at different chiral interfaces. From which it can be seen that the crack initiates at the joint position and then expands along the joint, finally the joined graphene fractures at the junction of two graphene.

Figure 5.22 shows the corresponding tensile stress-strain curve. It can be seen that the graphene with perfect joint presents a high tensile strength and failure strain, which is almost the same as the results of single piece of zigzag graphene (the slight difference is due to the deformation stress generated during the butt joining). However, when the graphene are joined at different chiral interfaces, the polygon defects produced at the joint will significantly reduce the tensile failure stress and damage strain of graphene. For example, in the forms of Z-Z joint and Z-A joint, the failure stress of joined graphene is reduced from 118.33 to 75.01 MPa. In addition, the joining of graphene with different chirality also results in a decrease in modulus of elasticity.

5.4.1.3 Effect of the Angle of Crystal Orientation on Mechanical Properties

The results in Fig. 5.22 show that the chirality of two pieces of graphene affects the quality of the graphene joint, thus affecting the performance of the two graphene. In fact, the difference of the chirality between the two pieces of graphene is the difference in their crystal orientation. In the actual process, the crystal orientation angle between the two graphene is random, which means that the connected graphene is composed of different crystal orientations, so the difference of the boundary defects after the connection of the graphene is different, which will inevitably influence the mechanical properties of the joint. In order to study the effect of crystal orientation on the mechanical behavior of graphene joints, the

5.4 Mechanical Properties of the Graphene Joint

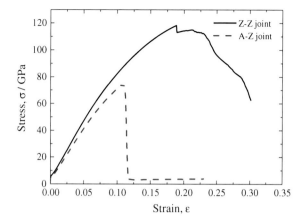

Fig. 5.22 Tensile stress-strain relationship of butt-joined graphene

graphene on the right side of Fig. 5.20a was rotated to obtain different crystal orientations. Figure 5.23 shows the effect of the crystal orientation on the performance of the bonded graphene joint, where the rotation of the right side of the graphene has a range from 10° to 90°. The tensile simulation conditions are consistent with the above.

Figure 5.24 shows the mechanical properties of graphene butt joints under different crystal orientations. First, the dynamic changes of the structure of joined graphene during stretching process can be obtained from Fig. 5.24a. Under the condition of uniaxial tension, the joined graphene displaces along the tensile direction, so the atomic stress in the graphene structure increases, and the stress concentration begins to appear at the butt joint position under the tensile strain. This is due to the presence of polygon and vacancy defects resulting in reduced cross

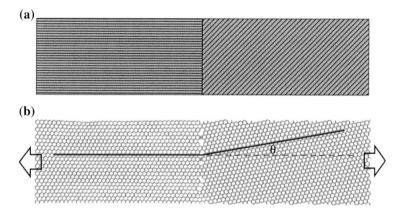

Fig. 5.23 Study model of the effect of crystal orientation on the mechanical properties of the butted graphene joint. **a** Schematic diagram of the research model, in which the angle between the horizontal line and the oblique line is the crystal orientation angle θ. **b** The corresponding MD stretching model

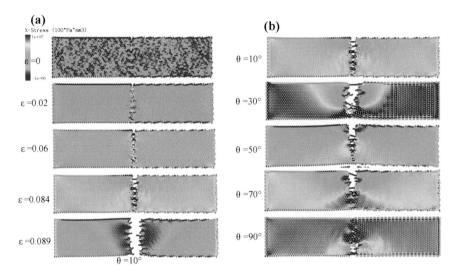

Fig. 5.24 Tensile stress contour diagram of butted graphene joint under different crystal orientations. **a** Dynamic change of the structure under tensile when $\theta = 10°$. **b** The tensile stress distribution before fracture of joined graphene under different crystal orientations

section at the joint position, so that the carbon-carbon bonds at the joint are subjected to greater stress. With the increase of tensile strain, the stress concentration at the graphene joint is further increased. Under certain strain conditions, the stress at the joint will be greater than the strength of graphene carbon-carbon bond, and then some of the graphene chemical bonds will be destructed. After the initial tensile defect is formed, as the stretching continues, the crack will expand along butt joint direction until the whole piece of graphene is destroyed. It is found that the failure process of joined graphene under different θ conditions follows the same law, that is, the stress concentration at the joint, the initiation of the defects, the expansion of the defects and the breakage of the graphene joint. Figure 5.24b shows the stress distribution of the graphene butt joint before fracture under different crystal orientations. It can be seen that the difference of the crystal orientation does not change the destruction position of the joined graphene during stretching process, but the stress distribution before fracture is different due to the difference of the crystal orientation. Meanwhile, it is also found that the fracture strain of joined graphene will be different due to the difference of the crystal orientation.

Figure 5.25 shows the effect of the crystal orientation angle on the strength and fracture strain of the joined graphene. At different crystal orientations, the variations of strength and failure strain obey a certain rule: when the crystal orientation angle is 0° and 60°, the failure stress and failure strain are the largest, which is close to the original monolayer graphene. When the crystal orientation angle is 30° and 90°, the strength and failure strain take the second place. When the crystal orientation angle is 10°, 20°, 40°, 50°, 70°, 80°, the strength and fracture strain of joined graphene

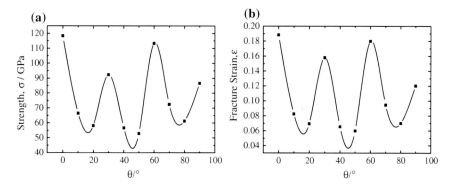

Fig. 5.25 Effect of crystal orientation angle θ on **a** strength and **b** failure strain of joined graphene

are small. We know that the graphene is composed of hexagonal honeycomb lattice, and its hexagonal regular unit makes the crystal structure of graphene have six rotational symmetry properties. Therefore, when fixing a piece of graphene and rotating another one, the six-time rotational symmetry characteristic allows the graphene to return to the initial structure every 60° rotation angle, thus forming a seamless and perfect joint with the first graphene. It is observed that the graphene can form the perfect butt joint at 0° and 60° crystal rotation angle. At the same time, the six-rotation symmetry also ensures that the structural unit of the graphene repeats at every 60° rotation angle. So the results of 30° rotation and 90° rotation should be substantially the same, and the results of 10° and 70°, 20° and 80° should also be substantially the same. In the case of 30° and 90°, the butt joint is actually the joined result between zigzag graphene and armchair graphene. Its strength and fracture strain are larger than the other cases (except for perfect joints), which indicates the joint generated between zigzag and armchair graphene has better mechanical properties.

5.4.2 Mechanical Properties of Overlapped Graphene Joint

5.4.2.1 Research Model

In the case of overlap, ion beam irradiation can promote the formation of chemical bonding between two layers of graphene, and the strong chemical bonds between layers can enhance the tensile strength of overlapped graphene. However, the irradiation of ion beam will also destroy the hexagonal crystal structure of graphene, which brings vacancy and polygon defects, and even makes the graphene structure in the irradiated area tend to be amorphous. The mechanical properties of the joined graphene are controlled by the number of chemical bonds and the degree of defects in the overlap region. It is actually the incident parameters of ions that

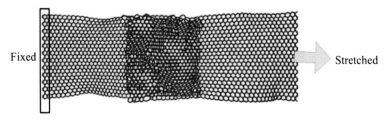

Fig. 5.26 Study model of mechanical properties of overlapped graphene joint. Reprinted from Ref. [15], Copyright 2014, with permission from Elsevier

determine the mechanical behavior. So the mechanical properties of overlapped graphene joined by ion beam irradiation were discussed based on the ion beam parameters. Figure 5.26 shows the mechanical properties of the graphene joint under the condition of overlap. Wherein the overlapped structure was derived from the result of ion beam irradiation. In this paper, the tensile properties of graphene joints were considered under different ion beam types (Ar and C) and ion beam parameters. Both the sizes of graphene are 50 × 100 Å and the boundaries are free boundaries. The left side of the graphene is fixed and the right boundary moves uniaxially at a strain rate of 0.0002 ps^{-1}. The interaction between graphene carbon atoms was described by AIREBO potential function. The calculation process was carried out by MD simulation software LAMMPS. Where the strength of the system is defined as the maximum stress during the stretching process and the failure strain is defined as the strain value corresponding to the strength. The system was fully relaxed at 300 K before stretching.

5.4.2.2 Mechanical Properties of Overlapped Graphene Under Ar Ion Irradiation

Ar ion cannot form a chemical bond with graphene C atoms, so under Ar ion irradiation, the formation of chemical bonds between graphene layers is due to the coordination of carbon atoms in graphene. In this section, the overlapped graphene joined under Ar ion beam irradiation with different ion parameters was stretched to research the mechanical behavior.

For the effect of the ion beam dose, the tensile strength of the graphene increases rapidly and then reaches the peak as the dose increases, and the corresponding ion beam dose at the peak is 1.9×10^{15} ions/cm^2; Then when the dose of the ion beam continues to increase, the damage stress and damage strain of graphene will decrease slowly. This is because that at the beginning, with the injection of ion beam, there will be new chemical bonds beginning to generate between graphene layers, and the number of new chemical bonds will gradually increase with the continuous injection of ion beam. However, the implantation of the ion beam also introduces defects in the graphene structure, which could deteriorate the mechanical properties of the joined graphene. The low dose of the ion beam cannot bring

5.4 Mechanical Properties of the Graphene Joint

enough chemical bonds, while excessive doses of the ion beam introduce too many defects in the graphene and cause serious amorphization in the irradiated area. Therefore, in the case of moderate dose, there will be a peak value for the strength and destructive strain of the overlapped graphene.

The same conclusion can be obtained for the effect of energy, that is, the failure strength and failure strain of graphene will increase rapidly with the increase of ion beam energy. After reaching the peak, the strength and failure strain of graphene will gradually decrease with the increase of energy. The peak value is at a position of 60 eV. This is due to the fact that the incident ion beam requires sufficient energy to strike the carbon atoms in the upper graphene, so as to overcome the initial energy barrier between the graphene layers to reach the bonding distance with the carbon atoms in the lower graphene. The ion beam with low energy cannot generate a strong enough collision with the carbon atoms in graphene to form chemical bonding. But at the same time, the irradiation of the ion beam will bring defects to the structure of graphene. The higher the energy, the greater the degree of defects caused by irradiation. So too high energy ion beam will strongly damage the upper and lower layers of graphene structure, which would greatly reduce the mechanical properties.

Thus, the optimal parameters for the joining of overlapped graphene under Ar ion beam irradiation can be obtained, where the optimum parameters are defined as the ion beam parameters corresponding to the maximum strength and maximum failure strain. In this case, the optimal parameters for overlapped graphene joined by Ar ion beam irradiation are 1.9×10^{15} ions/cm^2 and 60 eV. Under these optimal parameters, the strength of the graphene joint is 82.6 GPa and the failure strain is 0.19 [19].

5.4.2.3 Mechanical Properties of Graphene Joined Under C Ion Irradiation

For the carbon ion beam irradiation of graphene, because the carbon ion can be chemically bonded with the carbon atoms in graphene, the joining of graphene is induced by both the coordination of carbon atoms in graphene and "bridging" effect of embedded incident ions. The foregoing studies show that the "bridging" effect of the embedded atoms plays a major role in this joining process. Therefore, due to the different joining mechanisms for the carbon ion beam irradiation, the optimal parameters to generate a good joint between graphene layers under carbon ion beam irradiation will change.

Figure 5.27 shows the change of the strength of the graphene joint with the incident energy of carbon ions. The mechanical strength of the overlapped graphene joint increases rapidly with the increase of the energy of incident ion beam, and reaches the peak strength of 119.7 GPa at 40 eV. After 40 eV energy, the strength will gradually decrease will the increase of ion beam energy. Under the action of carbon ion beam, the mechanical properties of the graphene joint are close to that of the original graphene, and the corresponding value of the strength and fracture

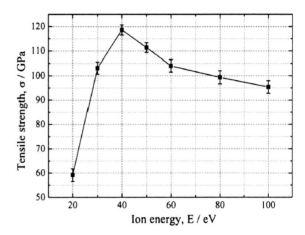

Fig. 5.27 The strength of overlapped graphene joint with different incident carbon ion beam energy. The ion beam dose is 1.06×10^{15} ions/cm^2. Reprinted from Ref. [15], Copyright 2014, with permission from Elsevier

strain is much higher than that of the Ar ion beam. For the effect of the dose, when the ion beam dose is too low, the number of embedded ions is too small, which would lead to a bad joint due to the overlap graphene is mainly joined by embedded ions. When the dose of ion beam is too large, even though there are enough embedded ions existed between the graphene layers to "bridge" the two graphene, the abundant ions will cause the amorphization in the graphene irradiation region, which could reduce the quality of the joint. It is found that a joint with good quality can be generated with the ion dose of 1.06×10^{15} ions/cm^2. For the influence of ion beam energy, since the embedded ion beam plays a major role in the joining of graphene, the incident ions require sufficient energy to penetrate the upper graphene and require a sufficient amount to chemically bond the upper and lower graphene. However, too much energy would make the incident ions penetrate the lower graphene, and the excess ion dose can cause repeated shocks to the graphene structure, resulting in excessive damage. Therefore, there are also optimal parameters existing for the joining of graphene by carbon ion irradiation, which are 1.06×10^{15} ions/cm^2, 40 eV, under which the strength of the joined graphene is 119.7 GPa. Due to the chemical bonding between the incident ions and carbon atoms in graphene, the optimal energy and dose for carbon ions are smaller than that of Ar ions.

When the incident parameters of the carbon ion beam were 1.06×10^{15} ions/cm^2 and 40 eV, it was found that the tensile failure of the bonded graphene existed in the original monolayer graphene, rather than the two graphene joint site (Fig. 5.28). This indicates that the quality of the graphene joint under the optimal parameters is very good, and the recombination process of the atoms at the joint counteracts the effect of the defects on the mechanical properties of graphene. However, if the energy of the ion beam is large (80 eV) or the dose is high (2.12×10^{15} ions/cm^2), there will be more defects produced in joint region. Under this condition, the tensile damage will initiate from the transition zone between the overlapped region and the monolayer graphene region. This indicates that

5.4 Mechanical Properties of the Graphene Joint

Fig. 5.28 Tensile morphology of overlapped graphene before failure joined with different parameters ion beam. Reprinted from Ref. [15], Copyright 2014, with permission from Elsevier

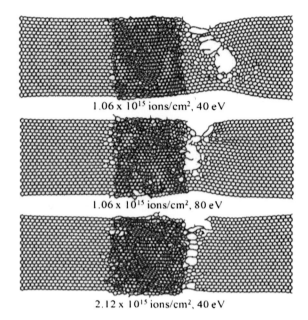

controlling the ion beam parameters not only affects the tensile strength of the whole graphene, but may also affect the destruction position of the graphene joint. For the weak transition region between the overlapping area and monolayer graphene, special attention needs to be payed in actual process.

5.4.3 Mechanical Properties of Butt Joint Constituted by Multi Pieces of Graphene

For the application of graphene, it is usually needed to join multiple graphene. The crystal orientation angle between different graphene is random, and there will be many polygon defects in the joint region due to the mismatch of the lattice. Since the crystal orientation of the adjacent graphene sheets is not the same, the polycrystalline graphene is formed after the joining. In this thesis, the polycrystalline graphene formed by multiple graphene sheets is stretched to investigate the tensile mechanical behavior, so as to study the mechanism of tensile failure of polycrystalline graphene.

Figure 5.29 shows the process of structural change of polycrystalline graphene under the condition of uniaxial tension, in which the contour distribution shows the atomic stress distribution, and the adjacent graphene sheets are connected with arbitrary crystal orientation. Before the uniaxial stretching, the atomic stress distribution of graphene is basically uniform, and there is no obvious stress concentration position. However, with the increase of tensile strain, stress concentration

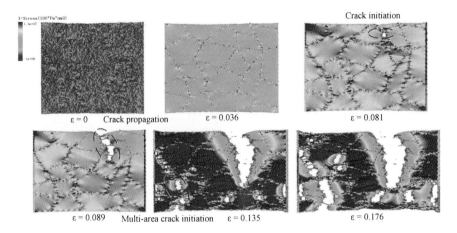

Fig. 5.29 Tensile properties of butt joint by multi pieces of graphene

begins to gradually appear in the grain boundary formed between the graphene sheets, which indicate that the joint is easy to become the concentrated region of the tensile stress. When the tensile strain continues to increase, the stress concentration in the grain boundary gradually increases and then exceeds the bonding strength between the carbon atoms of the graphene, which would lead to the formation of initial crack. As the strain further increases, the initial crack will gradually expand. The propagation path of the crack begins along the direction of the grain boundary (intergranular extension), then there is a phenomenon extending along the crystal (transgranular extension). Since the grain boundaries formed between the graphene sheets have polygonal defects and the ability to resist tensile damage is weak, it is easy to understand the intergranular extension. On the other hand, according to the above results, the tensile strength of graphene is related to the orientation angle between the graphene grain boundaries. Under certain orientation angles, the graphene joint will have a good tensile strength. Meanwhile, the tensile strength of graphene is anisotropy, that is, the tensile properties can be very different for different grained orientation. When the grain boundary of graphene is not perpendicular to the direction of tensile stress, the tensile force of the C–C bonds at the grain boundary will be further reduced, while the C–C bonds along the stretching direction will bear too much tensile load, which could exceed the strength of the interior C–C bonds, resulting in transcrystalline fracture. At this time, the tensile crack propagation direction is basically perpendicular to the tensile stress direction. However, when the bond strength between graphene grain boundaries is weak, the crack propagation will proceed along the grain boundary of graphene, so that the failure direction of graphene is not in the direction perpendicular to the tensile load. Therefore, for the graphene joint by multi pieces of graphene, the tensile load can result in intergranular fracture and transcrystalline fracture. In addition, in the process of stretching, crack initiation and expansion will happen simultaneously at different location of joined graphene, which means that different to the monocrystalline graphene, the polycrystalline graphene can have "dimple" phenomenon as that of the macroscopic ductile materials during stretching process, which presents a good toughness.

5.5 Electric Transport Properties of the Graphene Joint

The graphene joint is easy to become the weak position due to the defects generated during the joining process. In addition to the stress concentration under the external force, the defects at the joint will also affect the carrier transport properties of the graphene sheet. The electronic transport properties of graphene films are very important for their applications. The study of the influence of the joint on the electronic transport properties of graphene can further promote the significance of graphene joining. Also the applications of graphene can be expanded by controlling the properties of graphene with different types of joint. In this chapter, the electronic transport properties of graphene joint were calculated by DFT and NEGF. Overlap joint and butt joint were taken as reprehensive joints.

5.5.1 Electronic Transport Properties of Butt Joint

5.5.1.1 Research Model

In order to study the electronic transport properties of the butt joint, the graphene joined by the armchair type and zigzag type graphene was selected as the representative, and the transport properties under the condition of polygonal defects due to lattice mismatch were observed. The simulation model is shown in Fig. 5.30, in which the zigzag and armchair graphene are butt joined to generate a Z-A-Z sandwich structure. There are pentagon-heptagon defects pair existing in the joint due to the mismatch of the graphene lattice. The two edges of zigzag graphene are connected with infinite electrodes, forming the two electrodes model described in Chap. 2. The carbon atoms in the edges parallel to the transport direction are saturated with hydrogen atoms. A 15 Å vacuum layer is added beyond the hydrogen atoms to form a graphene nanoribbon structure. As for the width of the graphene nanoribbon, it is defined as number of zigzag bonds along the direction perpendicular to the transport direction, which is the same definition as literatures [20, 21]. So that the width of the zigzag graphene in this model is 6 (6-ZGNR), and the width of the armchair-type graphene is 11 (11-AGNR). The geometrical optimization of the system, the electron transport line and the calculation of current and voltage characteristics were done using the Transiesta software package described in Chap. 2. The GGA is taken as the exchange correlation function, and the cutoff energy of the plane wave is 150Ry. The K point grid of the Brillouin zone is $1 \times 1 \times 100$ (where Z is the electronic transport direction).

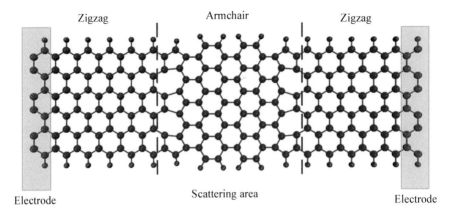

Fig. 5.30 Research model of the electronic transport properties of the butt joint (the red ball represents the hydrogen atom, the blue ball represents the carbon atom, the black dotted line is the regional divider)

5.5.1.2 Effect of Polygon Defects on the Transport Properties of Butt Joint

The mismatch of lattice at the joint tends to result in the formation of polygonal defects, which may have an effect on the carrier transport properties. The ideal zigzag graphene nanoribbon exhibits typical metal properties. The ideal armchair-type graphene nanoribbon exhibit different properties depending on their width, and when they have a width of N = 3n − 1, it can present the characteristic of semiconductor with a band gap [22]. Therefore, the 6Z-11A-6Z joint studied in this paper is essentially a joint between metal and semiconductor. Figure 5.31 shows the transport line of the joint under the condition of zero bias. As a comparison, this paper also calculated the transport lines of the pure zigzag-type and pure armchair-type graphene. In which, the width and length of the pure graphene nanoribbon is the same as the Z-A-Z joint. It is found that the transport capacity of Z-A-Z joint graphene is obviously lower than that of the original complete graphene. After the joining, there will be a weak transmission peak near the Fermi level, but the transmission of graphene is almost cut off in a certain energy interval (−1.5 to 1.7 eV). And then with the increase of electron energy, the transmission coefficient gradually increases. This indicates that although the pure armchair-type graphene and zigzag-type graphene have a large transmission coefficient at zero bias, the vacancy defects at the joint will greatly suppress the carrier transmission process. With the increase of carrier energy, more and more transmission channels will be opened, and then transmission capacity will gradually increase.

Figure 5.32 shows the current and voltage curve of Z-A-Z butt joint. The results show that when the bias voltage between the two electrodes is less than 2.5 V, the current value through the butt joint is very small, and there is also a slight negative differential resistance phenomenon; when the bias between the two electrodes is

5.5 Electric Transport Properties of the Graphene Joint

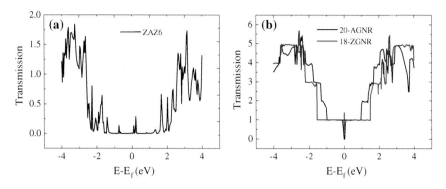

Fig. 5.31 Transport lines of **a** Z-A-Z butt joint and **b** pristine graphene under zero bias

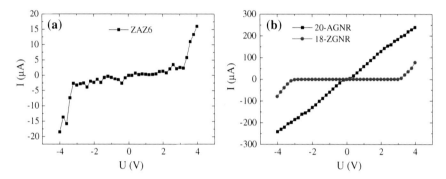

Fig. 5.32 The current and voltage relationship for **a** Z-A-Z butt joint and **b** pristine graphene under bias voltage

greater than 2.5 V, there will be more transport channels opened, so the current showed a significant increase trend with the voltage. Compared with the pristine graphene, it can be found that the current-voltage relationship of the joint is close to that of the zigzag graphene. It is conceivable that the transport properties of graphene may be controlled by adjusting the proportion of two different chiral graphene, which will be discussed later.

The above results indicate that the vacancy defects formed in the joint of different chiral graphene have a significant effect on the electronic transport properties. In order to explain the mechanism of the effect, the frontier molecular orbit is observed for the butt joint under zero bias, as shown in Fig. 5.33. It can be seen from the figure that for the pristine armchair graphene, the molecular orbit of the whole scattering region and the electrodes region are uniform, and the carrier has a strong transport capacity in the graphene plane. While for the zigzag graphene, the molecular orbit is mainly concentrated in the vicinity of the edge hydrogen atoms within the cutoff energy. So it has a strong local state and cut-off window for the transport properties. When a sandwich joint was formed by the armchair-type

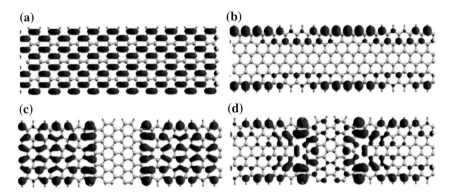

Fig. 5.33 Frontier molecular orbital of Z-A-Z butt joint at zero bias. **a** and **b** are the HOMO of armchair graphene and zigzag graphene, respectively, **c** and **d** are the HOMO and the LUMO of the Z-A-Z joint. The cutoff energy of the orbit is 0.009 eV, and the different colors represent the opposite sign of the orbit

graphene and zigzag-type graphene, it can be seen that the molecular orbital concentrates in the zigzag-type graphene region, indicating that the molecular orbital energy of the armchair-type graphene region is significantly lower than that of the zigzag graphene. And there is a serious local state in the polygonal defects located in the joint. These factors lead to the inhomogeneity of the molecular orbital in the plane of graphene joint, resulting in a significant reduction in the transport capacity of the carriers. It can be concluded that the transport capacity of the Z-A-Z butt joint graphene is controlled by two factors: one is the difference in the distribution of the molecular orbital energy levels in different graphene sheets, and the other one is the local state induced by polygonal defects. Since the molecular orbital level of monolayer graphene and the defect state after joining are closely related to its chiral direction (i.e., the orientation angle between the graphene), the electronic transport properties of graphene joint can be controlled by adjusting the crystal orientation angle between graphene sheets and the polygonal defects at the joint.

5.5.1.3 Effect of Joint Size on Transport Properties

The actual applications need to control the electronic transport properties of the joint. On the one hand, the crystal orientation angle between the two graphene can be adjusted, but it is very difficult to precisely control the deflection direction of graphene. On the other hand, the length of two graphene can be controlled to change the role of the each chiral graphene in the transport process, so as to achieve the purpose of controlling the transport properties.

In this work, the controlling of the transport properties is realized by changing the length of the intermediate armchair-type graphene. Three lengths of the middle armchair-type graphene were considered, i.e. 6, 10, 14 (the length is represented by the number of zigzag carbon chains). As the length increases, the proportion of

5.5 Electric Transport Properties of the Graphene Joint

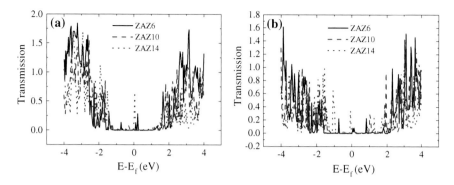

Fig. 5.34 Transport line of Z-A-Z joint under **a** 0 V bias and **b** 3 V bias

armchair-type graphene increases gradually in the joint. Figure 5.34 shows the transport lines of the butt joint under different lengths of armchair type graphene, where the results are extracted at 0 V bias and 3.0 V bias. Under different systems, the transport coefficient of the graphene joint near the Fermi level is always small, indicating that the defects in the joint part and the orbital energy difference between the graphene sheets will always produce a large block to the transport properties. However, the increase of the length of armchair-type part will lead to the appearance of the transmission peak near the Fermi level. Especially when the left and right electrode bias increases (3.0 V), there will be obvious transmission peak near the Fermi level. As a whole, the increase in the length of armchair-type graphene will increase the transport performance of the graphene joint.

Figure 5.35 shows the corresponding current-voltage relationship of the joint with different size armchair graphene. In the case of very low bias voltage, the current of the graphene joint is always low due to the low transmittance near the Fermi level. Before the 2.0 V bias, the current of the graphene joint is always below 5 μA, and there will be a slight negative differential resistance, and the increase in

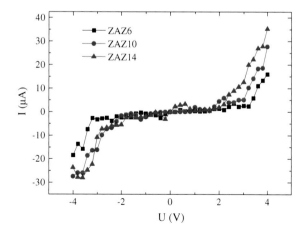

Fig. 5.35 The current and voltage relationship of Z-A-Z butt joint with different length of armchair-type graphene

the length of the armchair part will lead to the increase of negative differential resistance under low bias. When the voltage exceeds 2.0 V, the current of the ZAZ14 joint will increase rapidly with the increase of the bias voltage. The current of ZAZ6 type joint and the ZAZ10 joint will also increase rapidly when the bias voltage exceeds 2.8 V. In addition, it can be seen that the longer the length of the armchair graphene, the greater the current of the graphene joint, for example, when the bias is 3.0 V, the current under the three structures are: 2.62 μA (ZAZ6), 5.16 μA (ZAZ10), 12.63 μA (ZAZ14). This further demonstrates that the increase in the length of the armchair type graphene will enhance the transport capacity of the Z-A-Z type graphene butt joint under high bias conditions. Therefore, the electrical properties of the graphene butt joint can be controlled by adjusting the length ratio of the bonded graphene sheets.

5.5.2 Electronic Transport Properties of Overlap Joint

5.5.2.1 Research Model

For the overlap type, the bonding form of the graphene joint and the defects are different from the butt joint, which would lead to different electronic transport properties. The joint made up of two pieces of overlapped graphene is mainly controlled by two mechanisms: one is new chemical bonds resulting from the coordination of the carbon atoms in graphene. The other one is the "bridge" effect induced by embedded ions, which could form new chemical bonding with the upper and lower graphene sheet simultaneously. Under these two mechanisms, the electronic transport properties of graphene joint will have different results. In this paper, the electronic transport properties of the graphene joint with the overlap type were studied by using the model shown in Fig. 5.36, for which the joint induced by "coordination joining" and "embedded ions induced joining" mechanisms were considered. Both the graphene sheets shown in the figure are zigzag type and the dangling bonds at the edge are saturated with hydrogen atoms. The model uses three zigzag graphene, two of which are in a plane with a certain horizontal gap between them and another piece of graphene locates with a vertical spacing of 3 Å. The width of overlap region for each side is 4 (the number of zigzag carbon chains), and the graphene joining between the layers is formed by moving the carbon atoms at the upper and lower layers or by embedding the nitrogen atoms. In order to consider the influence of different incident parameters on the electronic transport properties of overlapped graphene, the number of chemical bonds between graphene layers was changed. For other calculation parameters, the same settings as butt joint model were used.

5.5 Electric Transport Properties of the Graphene Joint

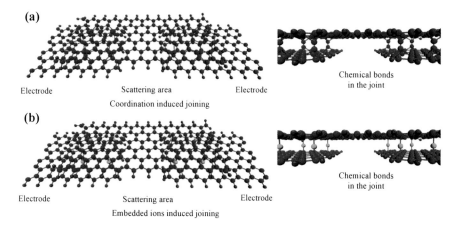

Fig. 5.36 Research model of the electronic transport properties of overlap joint. **a** Coordination induced joining. **b** Embedded ions induced joining. The red ball represents hydrogen atom, the blue ball represents carbon atom, and the green ball represents the embedded nitrogen atom

5.5.2.2 Electronic Transport Properties of Overlap Joints Under Different Joining Mechanisms

The graphene joined by different mechanisms will lead to the different electronic transport properties. Figure 5.37a shows the transport lines of the overlapped graphene under different joining mechanisms. It can be seen that for pristine overlapped graphene (before the formation of bonds), the transport coefficient of the carrier cross the graphene plane near the Fermi surface is zero, and there is a relatively large transport cutoff near the Fermi surface. While when the bonding is introduced by coordination or embedding bridge, the graphene plane will generate new transport channels near the Fermi surface, and new transport peaks will appear at different locations. These transport channels will increase the electronic transport

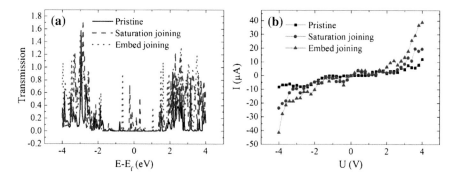

Fig. 5.37 a Transport line (zero bias) and **b** current-voltage relationship of graphene joint under different joining mechanisms. Reprinted with permission from Ref. [4]. Copyright 2017, AIP Publishing LLC

performance of overlap structure. Therefore, the chemical bonding between the graphene layers is important for the change of its transport properties, but the different bonding mechanisms have different effects on the transport properties of graphene. It can be seen from the figure, when the graphene layers are bonded through the coordination defects, there will be a new transport peak generated close to the Fermi level ($\sim$ −0.1, 0.3 eV), and the transport capacity in the vicinity of Fermi level increases significantly. When the layers are bonded through the embedded ions bridging, there will be new transport peaks generated at −0.8, 1.1 eV position. Figure 5.37b shows the current-voltage relationship of the graphene plane under different mechanisms. It can be seen that for the pristine overlapped graphene, due to the low transport capacity of the carrier, the current is always at a low value. While after the formation of chemical bonds between lower and upper layers, the current and voltage relationship will change. On the one hand, there will a negative differential resistance phenomenon existing. Although the negative differential resistance phenomenon is not obvious in the model of this work, it can be controlled by adjusting the bonding conditions between the upper and lower layers, and the differential resistance phenomenon is more obvious for the embedded ions induced joining. On the other hand, the current through the graphene plane will increase due to the formation of chemical bonding between the upper and lower layers under the same bias voltage, which is consistent to the increase of the carrier transport capacity under zero bias. In addition, when the embedded ions bridging is formed between the upper and lower layers, the increase of the current is more obvious than that of the coordination joining, especially when the bias is large. This indicates that when the bias is large, there will be more transport channels opened due to the newly formed chemical bonds.

In order to explain the increase in the transport capacity of graphene, the distribution of the orbital wave function of electrons in different cases is given by the front molecular orbital theory, as shown in Fig. 5.38. It can be seen from the figure that the chemical potential energy barrier between the original graphene layers is so high that most of the moving electrons are confined at the edge of the graphene. So there is no electron passing through the scattering region at the low bias voltage. Only under the higher bias, that there will be some transport channels opened. Therefore, for the pristine overlapped graphene, the capacity of electrons flowing from the left electrode to the right electrode is greatly reduced. While after the introduction of chemical bonding between the upper and lower layers, there will be some new local state generated in the bonding site. On the one hand, this local area will provide the transport channel of the electrons, which makes the electrons have higher possibility to transport through one electrode to another, on the other hand it will also block the electronic transport, inducing the formation of transport scattering. Compared to the original overlapped graphene, the overall effect of chemical bonding on the transport properties is promotion. When the bias at the graphene electrodes increases, there will be more new electron transport channels opened, then the promotion role of the chemical bonding on the electronic transport will be more obvious, resulting in the great increase of current through the graphene.

5.5 Electric Transport Properties of the Graphene Joint

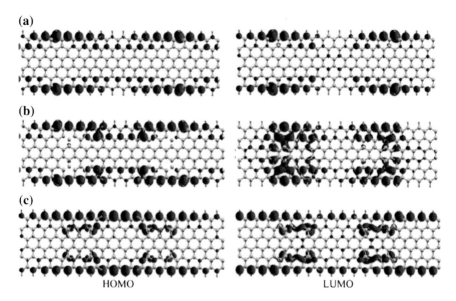

Fig. 5.38 The frontier molecular orbit of graphene joined by different mechanisms (zero bias). **a** The pristine overlapped graphene; **b** the coordination induced joining; **c** the embedded ions induced joining

5.5.2.3 Effect of the Number of Chemical Bonds on the Electronic Transport Properties

On the one hand, the newly formed chemical bonds of the overlap joint form new channels to promote electronic transport between the upper and lower layers of graphene. On the other hand, localized states are generated, which impede the electrons flowing between the left and right electrodes. When the number of chemical bonds formed between the upper and lower layers is changed, the transport properties of the joint will also be changed. Figure 5.39 shows the effect of the number of chemical bonds on the transport properties of overlapped graphene.

As can be seen from Fig. 5.39a, there will be new transport peaks appearing near the Fermi level at zero bias when the number of chemical bonds between the upper and lower layers increases. For example, when the number of bonds is 8, the peak of the transport line will appear at the energy of −0.8 and 1 eV. When the number of bonds is 12, the peak of the transport coefficient will appear at the energy 0 and 1.1 eV. So the transport channel of graphene is opened at certain locations. In general, the transport capacity of the joined graphene under zero bias tends to increase with the increase of the number of bonds. This is due to the fact that there is a relatively large energy barrier in the original overlapped graphene to prevent the passage of carriers. When the chemical bonds are formed, the transport probability between the upper and lower graphene is increased, and the increase in the number

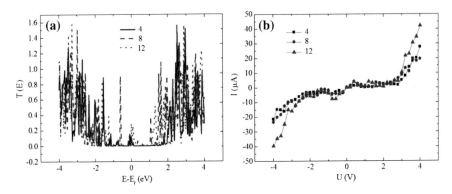

Fig. 5.39 Influence of the number of chemical bonds on **a** transport line under zero bias and **b** current-voltage relationship (coordination joining)

of chemical bonds will further promote the transport capacity. For the current and voltage relationship as shown in Fig. 5.39b, when the bias is small, there is almost no change for the current through the graphene plane under different number of chemical bonds. This is because under low bias, the transport capacity of graphene joint is weak, and the number of carrier transport channel is limited, thus the current under different number of chemical bonds is small. When the bias between the left and right electrodes is large (>2 V), more chemical bonding can form more transport channels, then the larger number of chemical bonds can promote the formation of a larger current. For example, when the voltage is 3.2 V, the corresponding currents are 7.93 μA (4), 13.15 μA (8) and 21.76 μA (12), respectively. This indicates that under the large bias, the number of chemical bonds has a more pronounced effect on the transport properties of overlapped graphene. Therefore, by regulating the energy and dose of the ion beam, we can control the chemical bonding of the upper and lower graphene, thus controlling the electronic transport properties of the joint. For the joint formed by embedded ions, a similar conclusion can be drawn.

5.6 Chapter Summary

In this chapter, the phenomenon and mechanism of graphene joining under the action of particle beam irradiation were analyzed by means of experiment and simulation. The mechanical and electronic transport properties of the joint were discussed. The main conclusions are as follows:

1. The action of ion beam irradiation, laser-current combined action and thermal annealing process can result in the joining of overlapped graphene. The joint formed by ion beam irradiation is due to the formation of new chemical bonds, which can be induced by the "coordination defects" of the carbon atoms in

graphene and the "embedded ions bridging" of the embedded atoms. The laser and thermal annealing action cannot generate the chemical bonding, and the joining phenomenon is formed only by strengthened intermolecular force between the graphene layers. Under the action of laser heat, the chemical bonds cannot be formed between two graphene with overlap and T-joint type, while the chemical bonds can be generated for the types of corner, butt and dislocated butt. This phenomenon is due to the coordination of the unsaturated edge carbon atoms.

2. Under tensile loading, the presence of defects in the joint makes it easy to become a tensile stress concentration area. For the butt joined graphene, the crystal orientation angle affects the presence of defects in the joint. There are two forms fracture modes for the joint made of multiple graphene: "intergranular fracture" and "transgranular fracture". Under carbon ions and argon ions irradiation, the optimal ion parameters to form a joint with the best mechanical properties are 1.06×10^{15} ions/cm^2, 40 eV and 1.9×10^{15} ions/cm^2, 60 eV, respectively.

3. Due to the difference of the orbital energy level between the different graphene sheets and the local state of the defects at the joint, the electronic transport capacity of the butt joined graphene is much weaker than that of the original monolayer graphene. When the overlap joint is formed, the new chemical bonds at the joint will promote the enhancement of the transport capacity. The promotion effect of "embedded ions joining" is stronger than that of "coordination joining". And the increase of the number of chemical bonds between the upper and lower graphene will further enhance the transport capacity.

References

1. Ye X, Huang T, Lin Z et al (2013) Lap joining of graphene flakes by current-assisted CO$_2$ laser irradiation. Carbon 61:329–335
2. Havener RW, Zhuang H, Brown L et al (2012) Angle resolved Raman imaging of inter layer rotations and interactions in twisted bilayer graphene. Nano Lett 12:3162
3. Kim K, Coh S, Tan LZ et al (2012) Raman spectroscopy study of rotated double-layer graphene: misorientation-angle dependence of electronic structure. Phys Rev Lett 108:246103
4. Wu X, Zhao HY, Pei JY, Yan D (2017) Joining of graphene flakes by low energy N ion beam irradiation. Appl Phys Lett 110:133102
5. Lin YC, Lu CC, Yeh CH et al (2012) Graphene annealing: how clean can it be? Nano Lett 12:414–419
6. Qi Z, Daniels C, Hong SJ et al (2015) Electronic transport of recrystallized freestanding graphene nanoribbons. Nano Lett 9:3510
7. Terrones M, Banhart F, Grobert N et al (2002) Molecular junctions by joining single-walled carbon nanotubes. Phys Rev Lett 89:075505
8. Krasheninnkov AV, Nordlund K, Keinonen J (2002) Ion-irradiaiton-induced welding of carbon nanotubes. Phys Rev B 66:245403
9. Li Y, Li B, Zhang H (2009) The computational design of junctions between carbon nanotubes and graphene nanoribbons. Nanotechnology 20:225202

10. Zou R, Zhang Z, Xu K (2012) A method for joining individual graphene sheets. Carbon 50:4965–4972
11. Wu X, Zhao HY, Zhong ML, Murakawa H, Tsukamoto M (2013) The formation of molecular junctions between graphene sheets. Mater Trans 54:940–946
12. Bao W, Miao F, Chen Z et al (2009) Controlled ripple texturing of suspended graphene and ultrathin graphite membranes. Nat Nanotech 4:562–566
13. Zakharchenko KV, Katsnelson MI, Fasolino A (2009) Finite temperature lattice properties of graphene beyond the quasiharmonic approximation. Phys Rev Lett 102:046808
14. Meyer JC, Geim AK, Katsnelson MI et al (2007) The structure of suspended graphene sheets. Nature 446:60–63
15. Wu X, Zhao HY, Zhong ML, Murakawa H, Tsukamoto M (2014) Molecular dynamics simulation of graphene sheets joining under ion beam irradiation. Carbon 66:31–38
16. Åhlgren E, Kotakoski J, Krasheninnikov A (2011) Atomistic simulations of the implantation of low-energy boron and nitrogen ions into graphene. Phys Rev B 83:115424
17. Kotakoski J, Jin CH, Lehtinen O et al (2010) Electron knock-on damage in hexagonal boron nitride monolayers. Phys Rev B 82:113404
18. Banhart F (1999) Irradiation effects in carbon nanostructures. Rep Prog Phys 62:1181–1221
19. Wu X, Zhao HY, Murakawa H (2014) The joining of graphene sheets under Ar ion beam irradiation. J Nanosci Nanotechnol 14:5697–5702
20. Son YW, Cohen ML, Louie SG (2006) Energy gaps in graphene nanoribbons. Phys Rev Lett 97:216803
21. Biel B, Blasé X, Triozon F et al (2009) anomalous doping effects on charge transport in graphene nanoribbons. Phys Rev Lett 102:096803
22. Topsakal M, Bagci V, Ciraci S (2010) Current-voltage (I-V) characteristics of armchair graphene nanoribbons under uniaxial strain. Phys Rev B 81:205437

Chapter 6
Fabrication of Graphene Nanopore by Particle Beam Irradiation and Its Properties

6.1 Introduction

With the implementation of the "Human Genome Project" and the "Thousand Yuan Genome" goal, nanopore sequencing, as a representative of the third generation gene detection technology, is expected to be a promising application in DNA sequencing, genetic testing, protein analysis and clinical diagnosis [1–3]. Nanopore detection technology is mainly limited by the diameter of nanopores and film thickness. Because most of the pore size and film thickness are much larger than the size of biomolecules, it can accommodate several to several tens of molecules at the same time, which would greatly reduce the signal sensitivity [4, 5]. Graphene is a hexagonal honeycomb lattice two-dimensional film formed by a single layer of sp^2 hybrid carbon atoms, with a single layer thickness of only 0.23 nm [6]. Due to the large specific surface area, good electrical properties and biocompatibility, graphene has become a hotspot for research as a new optional material of nanopore sensor [7, 8]. In addition, graphene nanopore has also been shown to have great potential applications in desalination [9], ion filtration [10] and so on. In order to realize the application value of graphene nanopore, it is necessary to explore the reliable processing methods and processing mechanism of nanopores. At the same time, the mechanical and electrical properties of nanoporous nanostructures are needed to be characterized to evaluate their special properties. Based on this, this chapter proposed the method of high energy focused ion beam and electron beam irradiation to process graphene nanopore, and the performance of nanopore was analyzed by computational simulation method.

6.2 Experimental Studies of Fabrication of Graphene Nanopore by Particle Beam Irradiation

6.2.1 Experiment Procedure

The experiment was carried out to process the nanopore in suspended and supported graphene. The experimental process includes:

1. Preparation of graphene sample

The graphene used in the FIB processing is monolayer prepared by CVD method. And the graphene used in the focused electron beam processing is multilayer (3–5 layer) prepared by CVD method. The preparation method is illustrated in Chap. 2. The number and uniformity of graphene were analyzed by optical microscopy and Raman spectroscopy. The results are shown in Fig. 6.1. Figure 6.1a shows that the prepared sample presents relatively large region showing

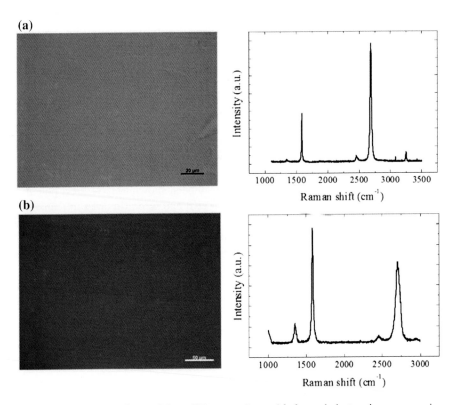

Fig. 6.1 Graphene samples used for **a** FIB processing and **b** focused electron beam processing (left view is the result of optical microscope, and the right view shows the Raman spectroscopic result)

6.2 Experimental Studies of Fabrication of Graphene Nanopore ...

monolayer continuous graphene, and the defects in the graphene structure are very few. Figure 6.1b shows that the prepared graphene exhibits a large area of continuous multi-layer graphene, with a small amount of defects in the graphene structure. Finally, monolayer graphene was transferred to a silicon substrate with a 300 nm oxide layer, and taken as ion beam processing sample, and the multilayer graphene film was transferred to a microwire copper mesh as an electron beam processed sample.

2. Processing of nanopore

The graphene was then processed using a FIB and an electron beam. First, the processing of the monolayer graphene nanopore structure on SiO_2/Si substrate was studied. The equipment is TESCAN FIB focusing ion instrument. The incident ions are 30 keV Ga^+, with ion probe current as 1 pA, and ion focal spot size is around 5 nm. The structure of the array pores is generated by controlling the running speed and residence time of the ion beam. Figure 6.2a shows the processing diagram. Then, the focused electron beam was used to process the suspended multi-layer graphene nanopore structure. The focused electron beam device is Tecnai G2 F20 S-Twin transmission electron microscope, and the acceleration voltage is 200 kV. The electron beam focusing diameter is within 1 nm. The processing diagram is shown in Fig. 6.2b.

3. Characterization of nanopore

The experimental results were characterized by AFM, Raman spectroscopy, SEM and TEM after the processing of the graphene nanopore structure by ion beam and electron beam irradiation.

6.2.2 Morphology Analysis of Graphene Nanopore

Based on the above experimental method, the graphene nanopore structure supported by the substrate was processed and the results were characterized by AFM. The results are shown in Fig. 6.3. The focused gallium ion beam can be processed

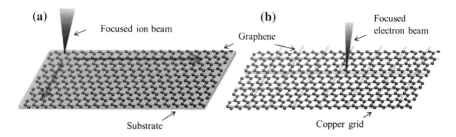

Fig. 6.2 Illustration of the nanapore processing in graphene by **a** FIB and **b** focused electron beam

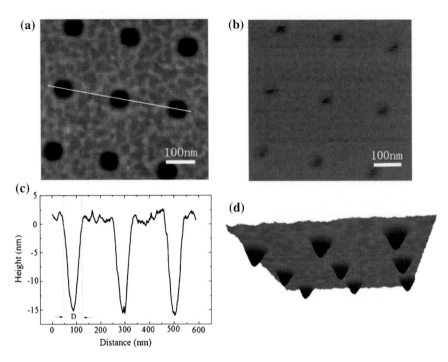

Fig. 6.3 Local AFM **a** topography image, **b** phase image and **c** profile, **d** 3D view of nanodot fabricated by FIB irradiation on supported monolayer graphene. The height profile in **c** corresponds to the line represented in **a**. Ion beam irradiation time is 0.8 μs, energy is 30 keV. Reprinted with permission from Ref. [15]. Copyright 2015, AIP Publishing LLC

to obtain a clear graphene nanopore structure. The periodicity of the array hole is about 200 nm, and the diameter of the array hole is about 75 nm. The holes are basically independent from each other, which indicates that the gallium ion beam irradiation can guarantee the integrity of the pore structure. The phase diagram (Fig. 6.3b) shows that gallium ions destroy the silica substrate while creating the graphene nanopore, so that the exposed second phase appears and the depth of destruction of the substrate is about 15 nm. The three-dimensional result (Fig. 6.3d) shows that after the ion beam processing, the hole structure formed in the substrate is an inverted cone shape, that is, the diameter of the hole in the silicon substrate gradually decreases along the irradiation direction. The depth of the pores obtained in this experiment is greater than the literature value [11], which is caused by two reasons: (1) The number of graphene layers used in Ref. [11] is not uniform, and there are some region presenting multilayer feature, so the existing of multilayer region lead to a higher impact resistance than that of monolayer graphene. At the same time, in the literature the graphene samples were grown directly on the SiC substrate, so the interaction between graphene and the substrate was relatively strong [12]. While the graphene samples used in this paper are basically monolayer, and graphene was transferred to the silicon substrate after being prepared.

The interaction force between the transferred graphene and silicon substrate is only a few meV [13], which is much less than that of the graphene grown on SiC. So the impact resistance of the samples in our experiment is greatly reduced. (2) Compared with the SiO_2 substrate, the SiC substrate has higher electron-hole formation energy (7.8–9 eV) [14]. It is more difficult to form a defective structure in SiC when subjected to a shock impact, thus resulting in a relatively small depth of the formed pores. The above results show that the substrate plays a very important role in the processing of graphene nanostructures.

In order to observe the possibility of nanopore processing in suspended graphene, focused electron beam was used to irradiate the suspended graphene on the copper mesh. Figure 6.4 shows the results of the electron beam processing of the suspended multilayered graphene. The graphene was irradiated with electron beam in TEM at high magnification for a certain time, and then the irradiated graphene structure was observed with a lower magnification. It can be seen that under the action of focused electron beam, the atomic motion of graphene is intensified, which is represented by the change of the brightness of the irradiated area. When the absorbed energy reaches the sputtering threshold of the carbon atom, the graphene structure will be locally destroyed. The damage originated from the electron beam irradiation area, and then extended to the surrounding. At last, there is a nanoscale pore (~ 7 nm) generated in graphene. But the structure is not stable, and the edge of the graphene nanopore will change, such as hole expansion, the fusion of defects, and then the structure is stabilized. For this size of graphene nanopore structure, it can be applied to the detection of biological macromolecule structures

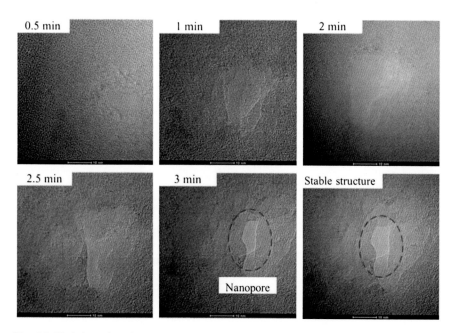

Fig. 6.4 Variation of graphene structure under focused electron beam irradiation (TEM)

such as protein. By further controlling the process parameters, it is expected that the nanopore structure with smaller size can be obtained by electron beam irradiation method. And this size of nanopore will be applied to DNA single molecule sequencing and other occasions. The mechanism of nanopore processing by focused electron beam, the dynamics variation of the atomic structure and the influencing factors will be described later by MD simulation.

6.2.3 Effect of Ion Beam Dose on the Properties of Graphene Nanopore

Particle beam with different parameters will have a great influence on the performance of graphene nanopore. The MD method will be used to study the effect of particle beam parameters on the morphology and properties of graphene. In this section, Raman spectra were used to characterize the performance of nanopore under different parameters of ion beam. This laid the foundation for further theoretical analysis.

Figure 6.5 shows the Raman characterization of the graphene nanopore structure under different doses of ion beam. The results show that the original graphene exhibits a good monolayer structure (2D peak is higher), and there is almost no defects for the structure (no D band). Ion beam irradiation will introduce defects to graphene (D band appears), and cause the reduction of the peak of 2D. The increase of the irradiation dose induced the continuous increase of the I_D/I_G ratio and continuous decrease of the peak of 2D, indicating that the graphene structure is continuously destroyed, and the graphene gradually changes to the nanocrystalline structure [16]. When the ion beam dose is particularly large (100 μs), graphene will be completely amorphous (2D band disappears). Therefore, as the ion beam dose increases, the graphene structure will have a transition process as defect appearance,

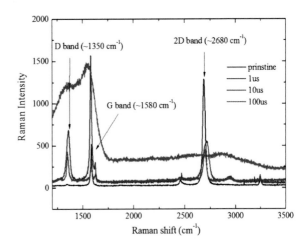

Fig. 6.5 Raman spectra of supported monolayer graphene after being irradiated with different dose ion. The dwell time of each incident ranges from 1 to 100 μs. Reprinted with permission from Ref. [15]. Copyright 2015, AIP Publishing LLC

nanocrystalline structure, and amorphous structure. According to the results in Fig. 6.3, it can be seen that the nanopore structure can be processed with the ion beam residence time of 0.8 μs. For the processing of graphene nanopore, it is necessary to control the ion beam parameters to obtain the nanopore structure with the desired characteristics.

6.2.4 Summary of the Experiment

In this section, the nanopore structure was processed on the supported monolayer graphene and the suspended multilayer graphene by using the FIB and the electron beam. Experiments show that the FIB can process a pore with tens of nanometers diameter. And by reasonably controlling the ion parameters or species, a pore with several nanometers may be obtained, for example, Abbas et al. [17] reported the use of focused He ions to achieve the 5 nm graphene nanoribbon. The focused electron beam can process smaller nanopores, and the literature [18, 19] also reported the use of focused electron beams to process 2–5 nm graphene pore structure. Future studies will focus on the pore size control and edge modification of graphene nanopore processed by particle beam irradiation. The subsequent simulations are based on the experimental results in this section to establish the corresponding theoretical model.

6.3 Theoretical Analysis of the Fabrication Mechanism of Graphene Nanopore

According to the results of the study in 6.2, and the conclusions of Refs. [17, 18], it can be seen that the use of FIB and electron beam can be used to process nano or even sub-nanometer level pores. While it is unclear how the performance of nanopore will variate with ion beam parameters and the graphene properties. Meanwhile, it is necessary to know the mechanism and microcosmic dynamic process of graphene nanopore processed by focusing ion beam and electron beam irradiation. Based on this, the classical MD simulation method was used in this section to study the focusing ion beam and electron beam processing of graphene nanopore.

6.3.1 Research Model

Figure 6.6 is the MD model. Wherein Fig. 6.6a shows the study model of electron beam processing of graphene nanopore. Figure 6.6b shows the study model of FIB

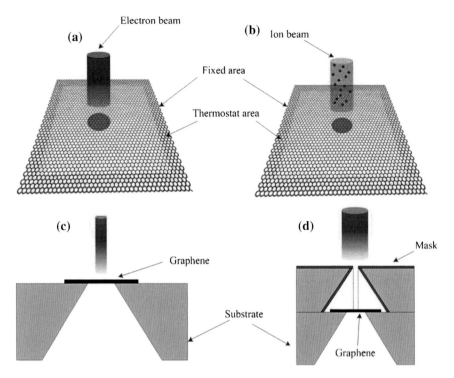

Fig. 6.6 Simulation model and the corresponding experimental schematic of graphene nanopore processed by **a, c** electron beam and **b, d** ion beam irradiation. Reprinted from Ref. [26], Copyright 2015, with permission from Elsevier

processing of graphene nanopore. Corresponding experimental illustrations are shown in (c) and (d). Figure 6.6 only shows the model of suspended monolayer graphene, the multi-layer graphene and substrate-supported models have similar simulation conditions and are therefore not shown separately. All MD simulations were carried out using LAMMPS software, in which the interaction between carbon atoms in graphene was described by AIREBO, and the cutoff radius of the potential function was 2.0 Å.

For the simulation of nanopore processing by FIB, the study was carried out using Ar ion. Ar ion and Ga ion have similar effects on the carbon atoms when incident on graphene (both do not form chemical bonds), so it can describe the same processing mechanism as Ga ion, and Ar ions are also easy to get in the experiment [20, 21]. The interactions between Ar ions and carbon atoms in graphene, Ar ions and substrates atoms were described by ZBL potential function. For the processing of the substrate-supported graphene nanopore structure, SiO_2 was used as the substrate, and the plane in contact with the graphene is the (0001) plane with the oxygen terminal. The initial distance between the graphene and the silica is 2.9 Å. The interaction between the atoms in the SiO_2 was described by the Tersoff potential function. In order to describe the cascade collision process, a short-range

6.3 Theoretical Analysis of the Fabrication Mechanism of Graphene Nanopore

repulsive force was added to the Tersoff potential function. The interaction between the carbon atoms in graphene and the silicon atoms in the substrate was represented by the Lennard-Jones (LJ) potential function. In which the function form is $V_{ij}(r) = 4\varepsilon_{ij}\left[(\sigma/r)^{12} - (\sigma/r)^{6}\right]$, where i = C, j = Si or O, r is the interaction distance, ε_{ij}, σ and the cutoff radius are consistent with the literature [22]. The graphene size is 10 × 10 nm, and the Ar ions are initially located in a cylindrical region of 40 Å above the graphene plane and then incident in a direction perpendicular to the graphene plane every 500 MD steps. The energy of the ion beam ranges 20–1000 eV, and the dose ranges 20–600 (the number of incident ions). By using these values, the effect of ion beam parameters on the structure and properties of graphene can be considered. Meanwhile the number of graphene layers is changed from single layer to five layers to consider the processing of nanopore in multi-layer graphene. The diameter of the irradiated cylindrical region is set to 2 nm, and the resulted nanopore structure can be used for DNA single molecule detection or desalination. For this simulation condition, the ion beam irradiation region can be controlled by the mask method [23] in the experiment (Fig. 6.6d). The same simulation conditions are used for the substrate-supported graphene nanopore processing. The thickness of the substrate is set to be 3.5 nm due to the limitation of the simulation time. The charge of the ions was not taken into account in the simulation.

The simulation method of electron beam irradiation is similar to that used in the collision researches of CNT [24, 25]. Wherein the irradiation process of the electron beam was mimicked by periodically assigning a velocity to the PKAs in graphene. For each collision, the PKA is randomly selected from the carbon atoms in the irradiated region shown in Fig. 6.6a, and the irradiated region has a diameter of 2.0 nm. A single PKA collision occurs every 10 MD steps, which lasts 2 ps, followed by 1 ps of structural relaxation. Then the next 2 ps collision and 1 ps relaxation are repeated. The whole simulation used a total of 20 times of such collision-relaxation cycles. The dose of the electron beam is expressed as the number of cycles, and the energy change of the electron beam is expressed as the energy assigned to PKAs, which ranges from 0.5 to 500 eV.

During the process of particle beam collision, the edge carbon atoms of graphene were fixed (Fig. 6.6a, b) to represent the actual fixed boundary conditions in Fig. 6.6c, d. In addition, a thermal bath was applied to the region near the edge of graphene to absorb the stress wave generated during the collision process. For that region, a Langevin hot bath was used to keep the temperature at 300 K. Before the collision process, the graphene structure was first equilibrated at room temperature for a sufficient period of time; then the ion beam or electron beam was incident to the graphene structure with different energy and dose, and the collision process was simulated in NVE ensemble. After the irradiation, the graphene structure was annealed at a high temperature of 2000 K (which can be used to simulate the actual low temperature annealing for a long time), so that the structure will not change any more; finally, the graphene structure was cooled to 300 K, and remained at that temperature for long time.

6.3.2 Processing Mechanism of Nanopore

Figure 6.7 shows the results of nanopore processing by FIB irradiation on single-layer suspended and supported graphene. The low energy and low dose ion beam incident will bring sputtering vacancy defects in graphene, and these defects increase with the increase of ion beam energy and dose. When the energy or dose of the ion beam increases to a certain extent, the defects will combine to generate larger size defects, resulting in the formation of nanoscale pore. For both the supported and suspended graphene, ion beam irradiation can form a nanopore with certain size. There are hanging bonds generated at the edge of nanopore, which are unsaturated and easy to adsorb other atoms to get saturated, so they are the ideal location for the chemical element doping of the graphene nanopore structure [9, 10]. When the

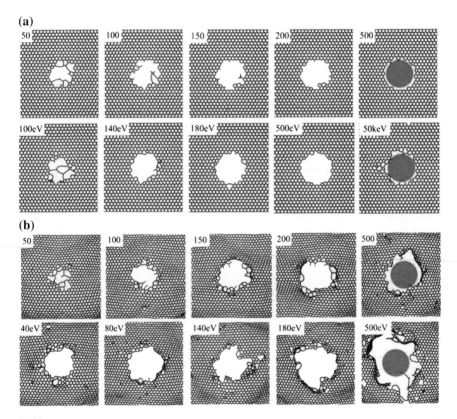

Fig. 6.7 Fabrication results of nanopore structure in **a** suspended and **b** supported graphene under ion beam irradiation. The influences of ion dose and energy are taken into consideration. For the influence of incident dose, the ion energy is set as 500, 80 eV for suspended and supported case, respectively. For the influence of incident energy, the ion dose is set as 400 for both suspended and supported case. The red region represents irradiated area of ion beam. Reprinted from Ref. [26], Copyright 2015, with permission from Elsevier

6.3 Theoretical Analysis of the Fabrication Mechanism of Graphene Nanopore

energy or dose of the incident ion beam is further increased, the roundness of the graphene nanopore (which describes the degree how the nanopore approaches a perfect circle, zero represents the complete circle) gradually approaches zero, at this time the desired nanopore structure is obtained. But when the incident ion energy is too large, the performance of graphene nanopore will be weakened, meanwhile too much energy and excessive dose will affect the equipment in the experiment, so it is necessary to discuss the optimal parameters (the lowest incident energy and dose to get the complete nanopore structure) for nanopore structure processed by particle beam, which will be described in detail in the next section. At the same time, it can be found that the presence of the substrate has an effect on the formation of graphene nanopore: on the one hand, the existing of substrate would lead to more atomic sputtering of the graphene under the same incident parameters. This is due to the secondary collision induced by the sputtering of substrate atoms (see Sect. 3.3 for detail explanation). On the other hand, the presence of the substrate will reduce the accuracy of the formed graphene nanopore structure, that is, the size of the nanopore will be greater than the region of ion beam irradiation. And the nanopore is more prone to exhibit irregular structure. Besides, the presence of the substrate can also result in vacancy defects in the graphene structure far away from the ion beam irradiated region.

In order to further explain the formation mechanism of nanopore, the dynamic variation process of the graphene structure under ion beam irradiation was observed (the results of electron beam irradiation are similar, which are not shown here), as shown in Fig. 6.8. It can be seen from the figure that the irradiation of the ion beam will form atomic sputtering in the suspended and substrate-supported graphene and lead to the formation of the nanopore structure. However, the presence or absence of the substrate can bring different phenomena: (1) the incident of the ion beam will

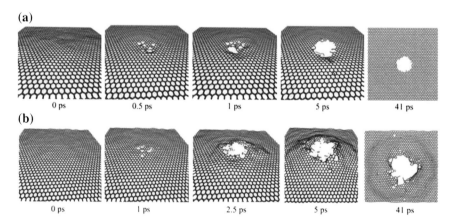

Fig. 6.8 Several snapshots of graphene during ion irradiation. **a** The suspended and **b** supported cases are indicated, individually. The irradiation energy is 300 eV, and ion dose is 400. The substrate is not shown for the supported case. Reprinted from Ref. [26], Copyright 2015, with permission from Elsevier

cause the impacted atoms to move along the incident direction, resulting in a depressed structure in the suspended graphene, as shown in Fig. 6.8a. The formation of the depression structure will bring the stress wave in graphene, and lead to the expansion of the depression area. However, under the condition of substrate support, the irradiation induced the formation of the convex structure, as shown in Fig. 6.8b. The convex structure also brings the transmission of the stress wave in graphene; (2) Under the same ion beam parameters, the performance of nanopore in suspended graphene structure is obviously better than that of supported graphene.

Figure 6.9 shows the schematic diagram of the atomic trajectories during the formation of nanopore in the suspended and supported graphene. It can be seen that for the suspended graphene, the carbon atoms will move rapidly along the incident direction after irradiated by ions, and then the carbon atoms near the collided atoms will also deviate from their original position due to the dragging action. So large area movement of carbon atoms will be gradually generated, which could explain the formation of depressed structure in Fig. 6.8a. When the energy of the incident ion beam is large, the carbon atoms in the graphene structure will be sputtered out, and the carbon atoms close to the sputtered atoms may break away from the graphene due to the "dragging" effect. At this time, the sputtered atoms are constituted by single atom or dimer, which is consistent with the incident sputtering process of nitrogen and boron ions [27]. The atoms around the sputtering region will then move in their equilibrium position and dangling bonds appear in these

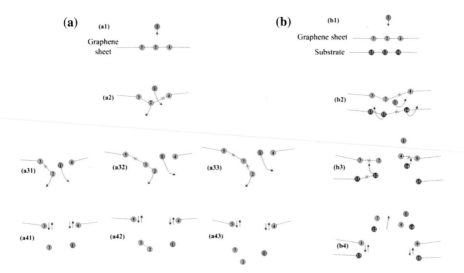

Fig. 6.9 Schematic plots of the atom trajectories for nanopore creation mechanisms in **a** suspended and **b** supported graphene. Atoms are represented by balls with number, where number 1 represents incident Ar ion, number smaller than 10 represents carbon atoms in graphene, number larger than 10 represents substrate atoms. The Ar ion, graphene atoms, and substrate atoms are also colored as green, grey and purple, respectively. The black solid lines represent the bonds between atoms, while black dash line means the bonds are broken. The red arrows represent the travel direction of atoms. Reprinted from Ref. [26], Copyright 2015, with permission from Elsevier

atoms. The above process shows that there are two mechanisms for the formation of nanopore structure in the suspended graphene under ion beam irradiation: "direct collision sputtering" and "nearby dragging". For the formation of nanopore in the supported graphene, the above two mechanisms also exist, while the presence of substrate will lead to the emergence of a third mechanism. As depicted in Fig. 6.9b, the incident ions will still have some residual energy after the collision with the graphene structure, which will bring the base atoms to be sputtered. Sputtering of the substrate atoms will produce a rebound to form a movement toward the graphene plane. In addition, the incident ions may also produce the sputtering bounce. At this time, the rebound of the base atoms and the incident ions leads to the collision between the rebounded atoms and graphene structure. So that the graphene plane structure is affected by the shock action with a direction opposite to the direction of the incident ions, which could result in the convex structure in Fig. 6.8b. When the energy of the rebounded atoms is large enough, they will have the energy to knock out the carbon atoms in graphene structure. The sputtering by the rebounded atoms is called "secondary sputtering". Therefore, the formation of nanopore structure in supported graphene is caused by the direct collision of the incident ions, the nearby dragging of the adjacent atoms, and the secondary collision of rebounded atoms. Since the movement of the base atoms and the rebounded ions is relatively random, there will be carbon atoms at the position far away from the irradiated region sputtered, which could explain the irregularity of the nanopore morphology and appearance of the vacancy defects near the pore structure for the processing of nanopore in supported graphene.

6.3.3 Influencing Factors of Nanopore Processing

It can be seen from Fig. 6.7 that the parameters of the incident particle beam will have a great influence on the structure of the graphene nanopore. In the experiment, on one hand it is necessary to obtain the complete structure of graphene nanopore. On the other hand, it is also needed to control the energy and dose of the irradiated particle beam to reduce the influence of excessive parameters on the equipment. At the same time, literature review has shown that for the DNA sequencing, nanopore in multilayered graphene structure has greater resistance to ionic current than that of monolayer graphene [28], resulting in a more easily distinguishable electrical signal. And the increase in the number of graphene layers can reduce the velocity of DNA molecule perforation [29], which is very important for DNA sequencing by using graphene nanopore. Therefore, it is very important to study the processing of nanopores in multi-layer graphene structure. The mechanism and influencing factors of multi-layer graphene nanopore processing were analyzed using the same method.

6.3.3.1 Influence of Particle Beam Energy and Dose

The morphology of the graphene nanopore can be represented by the number of sputtered carbon atoms. When the structure of the nanopore is stable, the number of sputtered carbon atoms will remain unchanged, at this time, the increase of the dose and energy will be detrimental to the process. Thus, the optimum parameters for the processing of graphene nanopore are defined as the minimum energy and dose required to get the stable number of sputtered carbon atoms.

Figure 6.10 shows the change of sputtering carbon atoms in the suspended and supported graphene under the action of FIB with different ion beam parameters. It can be seen from the figure that the number of sputtered carbon atoms increases with the energy and dose of the ion beam, which is due to the insufficient energy transferred to the graphene carbon atoms at lower energy. And also when the ion beam dose is low, there are not enough incident ions to make the carbon atoms in the graphene structure sputtering. With the increase of ion dose and energy, the

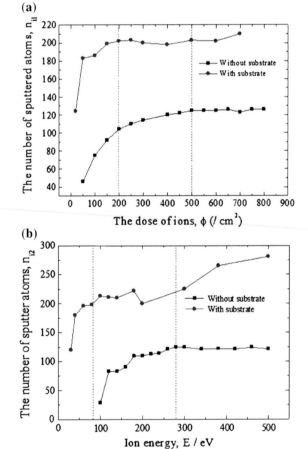

Fig. 6.10 Influences of incident parameters on the number of sputtered atoms under ion beam irradiation. **a** Influence of ion dose, ion energy is 500 and 80 eV for suspended and supported case, respectively. **b** Influence of ion energy, ion dose is 400 for both suspended and supported case. The dash lines are used to indicate the values defined as the optimal parameters. Reprinted from Ref. [26], Copyright 2015, with permission from Elsevier

6.3 Theoretical Analysis of the Fabrication Mechanism of Graphene Nanopore

number of sputtered atoms will gradually stabilize, which is due to the area of the irradiated region is certain. The stable moment corresponds to the time when the nanopore structure is completely formed. When the ion beam parameters continue to increase, the sputtering atoms in suspended graphene structure will not change any longer. While for supported graphene, due to the existence of the secondary collision from the sputtered base atoms, the number of sputtered carbon atoms will continue to increase to some extent, and there will be more and more vacancy defects formed around the nanopore structure with the increase of parameters. The optimum parameters for the processing of the graphene nanopore is defined as the minimum dose and energy corresponding to a complete nanopore structure, so the optimum energy and dose required for nanopore processing in the suspended and substrate-supported graphene are 280 eV, 500 and 80 eV, 200, respectively. It can be seen that the optimal energy and dose required to obtain a complete nanopore structure in the supported graphene are much smaller than that of the suspended graphene, which also indicates that the secondary collision process from the sputtered base atoms is the main factor for the formation of nanopore structure in supported graphene. In addition, although the energy of the ion beam is lower under the condition of substrate supported (80 eV vs. 500 eV), the number of sputtered carbon atoms is significantly larger than the suspended case.

For the case of focused electron beam irradiation, the collision process is described by the PKAs approximation, so the energy is defined as the energy absorbed by the carbon atoms and the dose is defined as the number of cycles to assign kinetic energy. Figure 6.11 shows the variation of the number of sputtered carbon atoms with the changes of electron beam irradiation energy and dose. It can be seen that the sputtering atoms increase rapidly with the increase of the incident energy until the maximum value is reached at 5 eV, and then the number of sputtered carbon atoms remains essentially unchanged when the electron beam irradiation energy continuously increases. This is because at low energy, electron beam does not have enough energy to make the carbon atoms in the graphene structure be sputtered off, and at high energy electron beam action, the carbon atoms in the irradiation area have been completely separated from the structure, which leads to a stable structure. The minimum energy required to achieve a stable structure is defined as the optimum energy, which is 5 eV. For the changes in dose, it can be seen that although the number of sputtered atoms increases as the dose increases, the degree of increase is small. So it can be thought that the dose has little effect on the structure of graphene nanopore if the energy of the electron beam is large enough. Meanwhile, the dose parameter is also barely considered in the actual application of electron microscopy equipment.

6.3.3.2 Influence of Graphene Layers

Compared with monolayer graphene, the nanopore structure in multilayered graphene will have different properties and different potential applications. However, the response of different layers of graphene under particle beam impact is different,

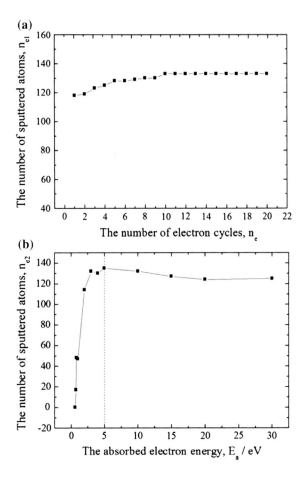

Fig. 6.11 Influences of incident parameters on the number of sputtered atoms under electron beam irradiation (suspended case). **a** Influence of the number of electron cycles, the absorbed electron energy is 3 eV. **b** Influence of absorbed electron beam energy, the number of electron cycles is 20. The dash line is used to indicate the value defined as the optimal parameter. Reprinted from Ref. [26], Copyright 2015, with permission from Elsevier

and the cascading collision of carbon atoms between layers could lead to a different phenomenon for the processing of multi-layer graphene.

Figure 6.12 shows the nanopore processing results in five-layer suspended graphene structure. It is seen that under the action of FIB, the nanopore structure with good quality can still be processed in the multi-layer graphene, but there will be more dangling bonds generated at the edge of the graphene. From the cross section and side view, it can be seen there are some chemical bonds formed between graphene layers under ion beam action. This is because that under the irradiation, carbon atoms in the upper layer graphene will be sputtered, resulting in the existence of the unsaturated carbon atoms, and meanwhile the cascade collision would also result in the sputtering in the lower layer graphene, leading to the formation of unsaturated carbon atoms in other graphene layers. Also, the collision will make the bonding type of carbon atoms transform from sp^2 to sp^3. Thus under the action of ion beam irradiation, there will be unsaturated carbon atoms generated in different layers of graphene, and the saturation of these atoms will induce the

6.3 Theoretical Analysis of the Fabrication Mechanism of Graphene Nanopore

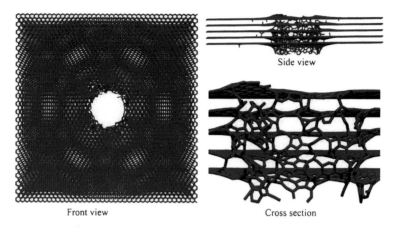

Fig. 6.12 Formation of nanopore in multilayer graphene (the number of layers is set as five) under ion irradiation. Reprinted from Ref. [26], Copyright 2015, with permission from Elsevier

formation of new chemical bonds between different layers of graphene. This phenomenon is also found in the joining of graphene sheets [30]. The newly generated chemical bonds can increase the interaction force between graphene layers, and reduce the possibility of the molecules entering the interlayer gap. So these chemical bonds are very important for the applications of multilayer graphene nanopore.

Figure 6.13 describes the optimal parameters for the processing of nanopore structure in graphene with different layers by ion beam irradiation. The optimal energy is defined as the minimum energy required to obtain a complete graphene nanopore structure. It can be seen that with increase of the number of graphene layers, the optimum energy required to obtain the nanopore structure is almost linearly increased. The inserted drawings give the structure of the nanopore under each energy. It is clearly seen that all the nanopore structures have excellent

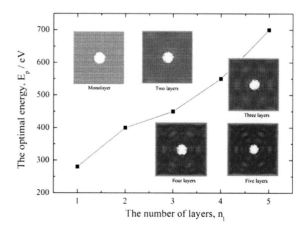

Fig. 6.13 Influence of the number of layers on the optimal energy for the fabrication of graphene nanopore under ion beam irradiation. Reprinted from Ref. [26], Copyright 2015, with permission from Elsevier

morphology. Meanwhile, it is found that the number of the dangling bonds at the edge will increase with the increase of graphene layers. These dangling bonds can block the perforation process of the biomolecules, and thus reduce the effective size of the nanopore. Besides, although the irradiation area of the ion beam in different layers of graphene is consistent, the size of the obtained graphene nanopore will be smaller with the increase of the graphene layer, which means that with the increase of the graphene layer, "nearby dragging" effect will be reduced.

6.4 Mechanical Properties of Graphene Nanopore

In practical applications, graphene nanopore often needs to sustain large stress and strain [9], so the mechanical properties of nanopore need to be analyzed. In this section, the uniaxial tensile mechanical behavior of nanopore was researched by MD method. The effects of nanopore size, graphene chirality and vacancy defects on the properties were discussed.

6.4.1 Research Model

Figure 6.14 shows the research model of the mechanical properties of graphene nanopore structure. The size of the graphene model is 20 × 20 nm, and the carbon atoms at one end of the graphene are fixed during the stretching process, and the carbon atoms at the other end are moving along the direction perpendicular to the boundary at a tensile rate of 0.001 ps^{-1}. Before stretching, the system is fully relaxed at room temperature and then deformed at a constant strain rate. After the tensile loading, the stress-strain relationship of the whole graphene and the atomic stress distribution in the graphene plane are extracted. For the calculation and

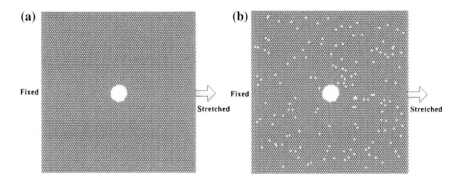

Fig. 6.14 Study models of mechanical properties of graphene nanopore. **a** Without defects. **b** With vacancy defects. Reprinted from Ref. [26], Copyright 2015, with permission from Elsevier

extraction methods, see Chap. 2. The strength and failure strain of the graphene nanopore structure are defined as the location where the maximum stress appears. The MD simulation software used in the calculation is LAMMPS, and the interaction potential between the carbon atoms in the graphene is AIREBO. The nanopore structure is formed by removing a certain amount of carbon atoms in the middle of the graphene sheet, and the size of the nanopore is represented by its diameter. Figure 6.14b shows the tensile model with vacancy defects in which the vacancy defects are introduced into the graphene nanopore structure in a randomly distributed form. The coverage rate of the vacancy defects is defined by the ratio between the number of missing carbon atoms to the total atoms in the system.

6.4.2 Tensile Failure Process of Graphene Nanopore

Firstly, the dynamic change of the structure of graphene nanopore under uniaxial tension is given, and the results are compared with those of the original graphene, as shown in Fig. 6.15. It can be seen from the figure that the original graphene can resist the large tensile deformation during the stretching process, the graphene structure has no obvious stress concentration under the tensile condition, and the local C–C bonds at a certain position break after the tensile strain reaches a certain degree. Then crack gradually expands along the initial failure location and eventually the graphene structure breaks down. After the introduction of the nanopore structure in the graphene, there will be obvious stress concentration at the edge of graphene nanopore under tension. A significant tensile stress concentration occurs near the axis perpendicular to the direction of stretching, and a significant compressive stress concentration occurs on the axis along the stretching direction. With the increase of tensile strain, the tensile stress of the nanopore edge atoms will

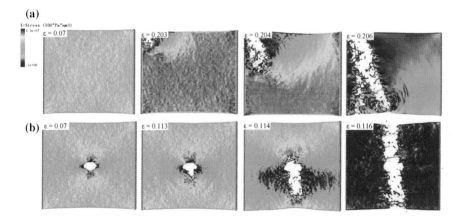

Fig. 6.15 Dynamic failure process of **a** pristine graphene and **b** graphene nanopore under the tensile load along zigzag direction

gradually reach the strength of the carbon-carbon bond of graphene, and then the initiation of the crack will occur. The initial crack will expand along the direction perpendicular to the tensile load, and eventually lead to breakage of the whole structure. Due to the presence of nanopore, the tensile damage strain of graphene is greatly reduced. It is reported that the stress concentration factor of nanopore in graphene structure is about 1.5–2.0 [31].

6.4.3 Effect of Graphene Chirality on Dynamic Failure Process

The tensile mechanical properties of the graphene structure are closely related to the direction of the tensile load. The strength and failure strain of the graphene structure in the zigzag direction are obviously larger than those in the armchair direction. When the nanopore is introduced into the graphene structure, what will the tensile behavior change with the chirality? Figure 6.16 shows the uniaxial tensile stress-strain relationship of graphene nanopores with different chirality, and the stress-strain relationship of the original graphene is shown as a comparison. The tensile properties of the original chiral graphene are consistent with those of the literature [32, 33], which illustrates the reliability of the method. When the nanopore structure is introduced, the tensile failure stress and the failure strain of the graphene are degraded largely, which indicates that the existence of nanopores will weaken the mechanical properties of graphene. But the elastic modulus of the graphene nanopore is basically the same as the original graphene. At this time, the tensile failure strain of different chiral graphene is the same, with a value of 0.12. The difference of their tensile strength is also relatively small, which is 73.9 GPa (zigzag type) and 67.3 GPa (armchair type), respectively. This indicates that the tensile mechanical properties of graphene with different chirality are becoming

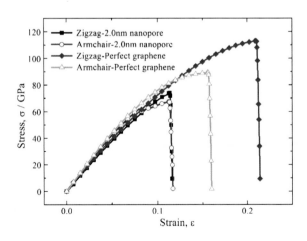

Fig. 6.16 Uniaxial tensile stress-strain relationship of graphene nanopore with different chirality. The size of nanopore is 2.0 nm. The chirality refers to the direction of the applied tensile loading. Reprinted from Ref. [26], Copyright 2015, with permission from Elsevier

6.4 Mechanical Properties of Graphene Nanopore

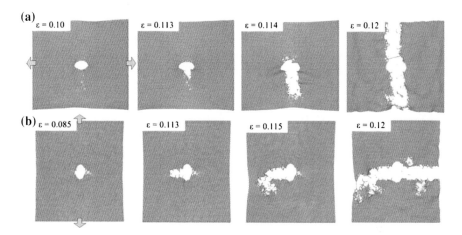

Fig. 6.17 Fracture process of **a** zigzag and **b** armchair graphene nanopore under uniaxial tension

close to each other due to the presence of nanopore. It is known that during the stretching process, the special structure of graphene makes the carbon-carbon bonds in the armchair direction bear relatively large tensile stress, while the carbon-carbon bonds along the zigzag direction only bear part of the tensile stress due to the certain angle between the C–C bonds and stretching direction. At this time the C–C bonds along the armchair direction is easier to be broken under the same tensile load. However, when the nanopore structure is introduced, the nanopore becomes the weak region of the graphene structure, and the effect of the nanopore structure on the stress distribution of graphene is greater than that of the graphene chirality. Under this condition, the destructive behavior of graphene is dominated by the nanopore structure, so that the different chiral graphene nanopore will present a similar fracture behavior in the stretching process under the condition of 2.0 nm nanopore.

Figure 6.17 shows the dynamic failure process of graphene nanopore with different chirality. It can be seen that the cracks are initiated at the edge of the graphene nanopore during the stretching process, and the initial vacancy defects are gradually fused to form a collective fracture of the chemical bonds. Then the crack extends along the direction perpendicular to the tensile load. The final fracture cross section is substantially perpendicular to tensile load direction. Although the initial crack formation of the armchair-type graphene is earlier than that of the zigzag graphene during the stretching process, the crack growth rate of the armchair-type graphene nanopore is significantly smaller than that of the zigzag graphene. Finally the different chiral graphene was destroyed under the same tensile strain ($\varepsilon = 0.12$).

6.4.4 Effect of Nanopore Size on Mechanical Properties

Figure 6.18 shows the change of the breaking strength of graphene nanopore with the nanopore size. It can be seen that the tensile failure strength of graphene nanopore decreases with the increase of nanopore size for different chiral graphene structures, and the tensile strength of armchair-type graphene nanopore is always lower than that of zigzag-type graphene. This is due to the stress concentration introduced by the nanopore structure. When the diameter of the nanopore increases, the number of carbon-carbon bonds in the tensile cross-section will be less, which could make the structure easier to be fractured under the same load condition. Although the effect of the nanopore structure on mechanical properties is greater than the effect of graphene chirality, there is still the effect from chirality. Therefore, the mechanical properties of graphene nanopore are determined by the combined effect from the nanopore structure and the chirality. However, the difference of the breaking strength of graphene nanopores with different chirality is always less than that of the original graphene with different chirality.

Figure 6.19 shows the atomic stress distribution of the graphene structure under tensile condition with different nanopore sizes. It can be seen from the figure that with different nanopore size, there will be stress concentration in the vicinity of the nanopore under the tensile condition. Then the crack will initiate due to the stress concentration, and then the crack will expand along the direction perpendicular to the tensile load, which could result in rapid damage of the structure. Meanwhile, with a larger nanopore diameter, the graphene structure will be destroyed at lower strain conditions (0.98 vs. 0.113). In Ref. [31], it is also reported that the stress concentration coefficient near the nanopore will increase with the increase of the pore size.

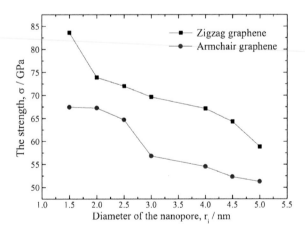

Fig. 6.18 The dependence of tensile failure strength of the graphene nanopore on the nanopore diameter

6.4 Mechanical Properties of Graphene Nanopore 165

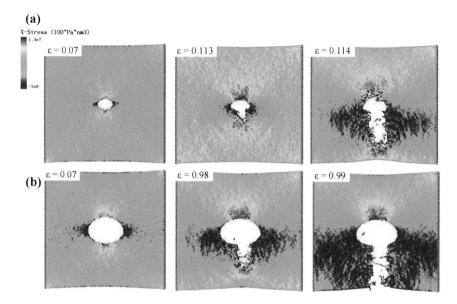

Fig. 6.19 Dynamic changes of the graphene nanopore under stretching. **a** Nanopore diameter is 2.0 nm. **b** Nanopore diameter is 5.0 nm

6.4.5 Effect of Vacancy Defect on Mechanical Properties of Nanopore

According to the results of Sect. 6.3, it can be seen that the sputtering of the base atoms during the nanopore structure fabrication will bring vacancy defects near the nanopore. Meanwhile, the preparation process will inevitably bring some defects to the graphene structure, and the main defect form is vacancy defect. Therefore, the graphene nanopore structure should be always accompanied by the vacancy defects, it is very important to understand the effect of vacancy defects on the mechanical properties of graphene nanopore.

In this section, the mechanical behavior of the graphene nanopore-vacancy structure under uniaxial tension was analyzed using the model shown in Fig. 6.14b. The conditions of the tensile simulation are consistent as described above. Figure 6.20 shows the tensile stress-strain relationship of graphene nanopore with different vacancy defect concentrations. It can be seen from the figure that when the vacancy defect concentration is less than 0.5% (0.5% corresponds to the absence of 73 carbon atoms), the existence of vacancy has little effect on the tensile properties of graphene. When the vacancy defect concentration is more than 1%, the failure strength of the pore structure is greatly reduced (e.g., the tensile strength of the

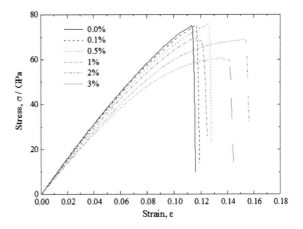

Fig. 6.20 Tensile stress-strain relationship of graphene nanopore with different concentrations of vacancy defect

graphene nanopore decreases by about 20% when the vacancy defect concentration is 3%). For the failure strain, it can be seen that the failure strain of the graphene nanopore will increase due to the presence of the vacancy. This is because the defect structure in the graphene leads to the migration of atoms and reorganization of the structure during the stretching process, which could result in the rotation of C–C bonds and a greater deformation of graphene structure without damage. It is consistent with the result of the influence of vacancy defects on the tensile failure strain of pristine graphene [34]. Meanwhile, the existence of vacancy defects will reduce the elastic modulus of the graphene structure, and the degree of decrease will increase with the increase of the vacancy defect rate.

For pristine graphene, a low concentration of vacancy defects (less than 0.086%) reduces the tensile strength of the graphene structure by about 20% due to the presence of significant stress concentration induced by vacancy defects [35, 36]. When the nanopore is introduced, the graphene structure appears to be immune to the vacancy defects with low concentration. In order to explain this phenomenon, this section gives the stress distribution of graphene nanopore with different vacancy defect concentrations, as shown in Fig. 6.21. The tensile stress distribution of the graphene nanopore without vacancy defects is consistent with that reported in the literature [31]. For which the stress concentration occurs around the nanopore. When the vacancy defects appear, the stress distribution in the graphene structure will change. It can be seen that the stress concentration is gradually occurring near the vacancy defect, and the stress concentration around the nanopore will decrease. For the low concentration ($\sim$1%) vacancy defects (Fig. 6.21b), the stress concentration around the nanopore is greater than that near the vacancy defects, and the initial crack in the graphene nanopore will still be generated near the pore. While when the concentration of the vacancy defect is relatively large ($\sim$5%) (Fig. 6.21c), the stress concentration in the vicinity of the nanopore will be similar to that near the vacancy defect, which could lead to the initiation of crack simultaneously in the vicinity of vacancy defects and the nanopore. Therefore, in the case of low concentration, the existence of vacancy defects does not change the degree

6.4 Mechanical Properties of Graphene Nanopore

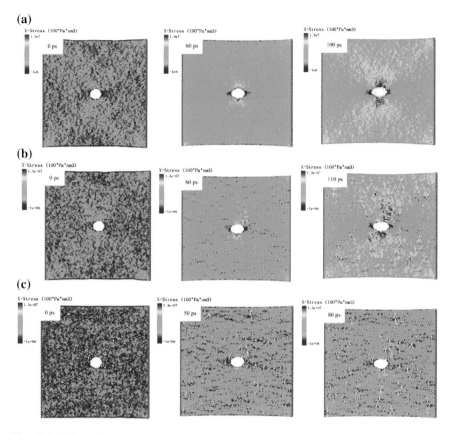

Fig. 6.21 Effect of vacancy defect on stress distribution of graphene nanopore. **a** The vacancy defect rate is 0%; **b** the vacancy defect rate is 1%; **c** the vacancy defect rate is 5%

and position of the main stress concentration of the graphene nanopore, and the tensile failure behavior of the graphene nanopore will not be changed obviously. While under the high concentration vacancy defects, the stress concentration coefficient and the concentrated position of the graphene structure will change, so at this time the tensile mechanical behavior of the graphene structure is determined by both the nanopore structure and the vacancy defect. So, the graphene structure will be mechanically immune to the low concentration of vacancies. This defect "immune" phenomenon also exists in the structure of other nano-materials, and has been theoretically and experimentally confirmed [37, 38].

Figure 6.22 describes the dependence of mechanical behavior on vacancies for NPG with different radius of nanopore. As displayed in Fig. 6.22a, when the coverage of vacancies is low (small than 3%), the strength of NPG is determined by the concentration of stress near the nanopore, so the NPG with bigger nanopore is easier to be destroyed. With the increase of coverage rate, the role of vacancies in the strength is increasingly important, hence at that coverage (3–4%), the strength

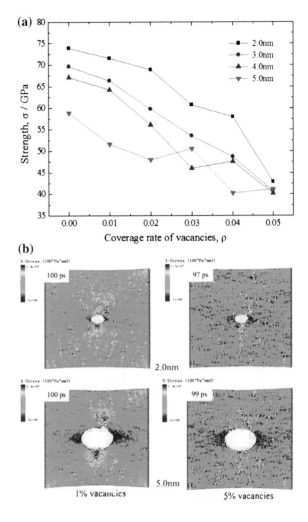

Fig. 6.22 Influence of vacancies on the mechanical properties of NPG with different radius of nanopore. **a** Strength variations of NPG with different coverage rate of vacancies **b** corresponding stress distribution of NPG with different radius of nanopore. Typical snapshots of NPG configure before failure are given

of NPG shows no solo-dependence on the radius of naopore anymore. In fact, at this coverage, there exists a competition for the influence on strength between nanopore and vacancies, and the winner is decided by both the nanopore radius and vacancies coverage. If we continuously increase the coverage rate of vacancies, the role of vacancies in the strength overrides pore structure, so the strength of NPG is determined by vacancies and different radius present almost the same properties. The corresponding stress distributions of NPG with different radius of nanopore depicted in Fig. 6.22b also show that for different radius of nanopore, the stress are still mainly concentrated around the nanopore for low coverage rate of vacancies, while the stress concentration happens both near the nanopore and vacancies for high rate of vacancies, which again confirms the nanopore structure determined strength and vacancies determined strength of NPG under different coverage rate of vacancies. We can also speculate in literature [37, 38] that though the

6.4 Mechanical Properties of Graphene Nanopore   169

nanocrystalline materials are insensitive to the flaw under small flaw size due to the grain boundary effect, the strength of nanocrystalline materials will be decided mainly by flaw if the flaw size is large enough.

6.5 Electronic Transport Properties of Graphene Nanopore

Nanopore can open the zero band gap structure of graphene, and make it have a relatively high switching ratio coefficient, thus graphene nanopore has a very important application potential in the graphene-based electronic components. The study of the electrical properties of graphene nanopore is very important for its application in electronic components. In this section, the electronic transport properties of the graphene nanopore structure were described by the combination of DFT and NEGF theory. The influence of nanopore size, nanopore shape and the presence of defects in nanopore were discussed.

6.5.1 Research Model

Figure 6.23 shows the model for the calculation of the transport properties of graphene nanopore. The study model is a ZGNR with $N = 20$ (N is the number of zigzag type carbon chain in the width direction). The green shaded portion represents the left and right electrodes, which are essentially semi-infinite graphene nanoribbons, and the middle region is the scattering zone. In the process of modeling, the carbon atoms in the middle of the scattering zone are deleted to form the graphene nanoribbon with the pore structure, and the center of the hole has the same distance to the upper and lower edges. The effect of the size and shape of nanopore on the transport properties were studied by controlling the shape of the pore and the number of deleted carbon atoms. The diameter of the nanopore shown in the figure is 10.24 Å, and the shape of the nanopore is a regular hexagon (rounded after hydrogenation). The number of carbon atoms in the whole transport system is about 620 (this number will change slightly for different systems) after the carbon atoms in the pore structure are removed, which is large enough be comparable with the experimental system. For the upper and lower edges of the model and the edge of the hole, hydrogenation is used to saturate the edge dangling bonds to make the structure more stable. At the same time, a number of single vacancy defects (position 1–8 shown in the figure) were randomly introduced to simulate the effect of the vacancy defect caused by the nanopore fabrication process on the transport properties. The effect of the concentration of defects (the number of missing carbon atoms) on the properties of graphene nanopores was discussed.

Fig. 6.23 Study model of electronic transport properties of graphene nanopore. The blue ball represents the carbon atom, the red ball represents the hydrogen atom, and the numbered ball represents the missing atomic position in the process of making the vacancy defect

Left Lead   Scattering Region   Right Lead

In this paper, the geometrical optimization of the system, the electron transport line and the current and voltage characteristics were calculated using the Transiesta software package described in Chap. 2. The GGA was used as the exchange correlation function. The cutoff energy of the plane wave was 150 Ry, and the K point grid of the Brillouin zone was $1 \times 1 \times 100$ (where Z is the electron transport direction). The transport properties of the perfect graphene were also calculated as a comparison.

6.5.2 Effect of Nanopore Size on Electrical Properties

Figure 6.24a shows the electronic transport lines of graphene with different size nanopore (zero bias). It can be seen from the figure that the original zigzag graphene is metallic, but the electronic transport properties of graphene are decreased after the introduction of nanopore defect. This indicates that the nanopore will inhibit the transport of carriers, and the reason will be explained later by frontier molecular orbital theory. The inhibitory effect will be enhanced with the increase of nanopore size. It can be seen from the figure that when the size of the nanopore is 15.03 Å, the transmittance near the Fermi level is close to zero, and the transport channel of the graphene near the Fermi level is completely blocked, which indicates

6.5 Electronic Transport Properties of Graphene Nanopore

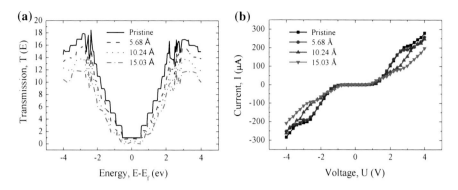

Fig. 6.24 a Electronic transport lines (zero bias) and b current-voltage relationships for different sizes of nanopore

that the graphene nanopore exhibits the properties of semiconductor. This also shows that the structure of nanopore can be used to open the zero band structure of graphene. Figure 6.24b shows the current and voltage curves of graphene with different nanopore sizes. It can be seen from the figure that the introduction of nanopore will suppress the carrier transport of graphene, and the transport current will decrease under the same voltage. The current flowing through the graphene structure is small at low voltage, and the effect of the nanopore on the transport current is also small. As the voltage increases, more and more transport channels are opened, and the influence of nanopore on the transport current will increase significantly. It can be seen that the effect of graphene nanopore on the current is small when nanopore is small (5.68 Å). With the increase of nanopore size, the effect of the graphene nanopore on the current is increasing. For example, when the bias voltage applied across the electrodes is 3 V, the current through the 15.03 Å graphene nanopore is reduced to be half of the original graphene, and the currents at different sizes of nanopore are 209.83 μA (original) 190.04 μA (5.68 Å), 153.56 μA (10.24 Å), 102.56 μA (15.03 Å).

In order to explain the mechanism of the effect of nanopore defect on the electronic transport properties, we plot the molecular projection self-consistent Hamiltonian of graphene nanopore at zero bias, as shown in Fig. 6.25. It can be seen from Fig. 6.25a that the HOMO states of the original graphene and graphene nanopore are almost the same at zero bias, and both the states are mainly localized at the edge atoms. For the lowest occupied state (LUMO), it is also found to have the same distribution. Therefore, the nanopore structure has little effect on the frontier molecular orbital distribution of the graphene nanoribbon. Figure 6.25b shows the distribution of wave function of the non-frontier molecular orbital (LUMO+4). It can be seen that the molecular orbital will appear an obvious local state in the vicinity of the nanopore. Due to this localized state caused by the nanopore, the carrier transport through the graphene plane will be suppressed.

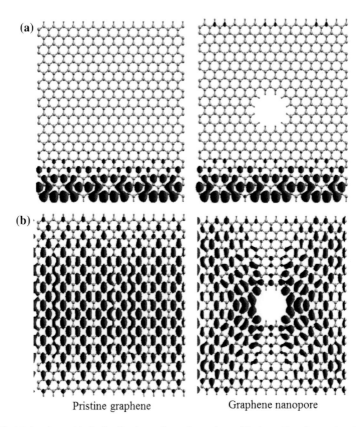

Fig. 6.25 Molecular orbital distribution of graphene (zero bias). **a** Frontier molecular orbital (HOMO); **b** Non-frontier molecular orbital (LUMO+4). The orbital isoplane is 0.005 eV, and the different colors represent positive and negative molecular orbital

Therefore, the localized state of non-frontier molecular orbital caused by nanopore structure is the main reason for the decline of graphene transport performance.

6.5.3 Effect of Vacancy Defect on Electrical Performance

The results of Sects. 6.2 and 6.3 show that fabrication of graphene nanopore will inevitably introduce defects into graphene, and there are also some inherent defects in the graphene structure, which are mainly in the form of vacancies. So for the actual situation, the nanopore structure and the defects are coexisting. Therefore, it is necessary to analyze the transport properties of graphene with combined nanopore-defects structure, as shown in Fig. 6.23.

Figure 6.26 shows the transport lines of the graphene nanopore with different concentrations of vacancy defects. It can be seen from the figure that the transport

6.5 Electronic Transport Properties of Graphene Nanopore

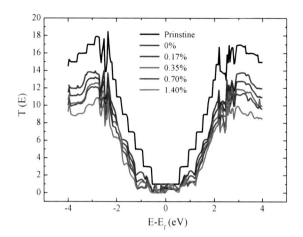

Fig. 6.26 Electronic transport lines of graphene nanopore with different concentrations of vacancy defect (zero bias). The nanopore diameter is 10.24 Å

capacity of the graphene nanopore will be further reduced with the presence of defects, indicating that the defects will further hinder the transport of the carriers in the graphene. It is due to the fact that the defects can cause the localized state of the graphene structure. However, the local state caused by the nanopore structure is stronger than the vacancy defects, so the effect of the nanopore on the transport capacity is obviously larger than that of the low concentration defects. With the increase of the defect concentration, the effect of the vacancy defect on the transport capacity is becoming more and more obvious. When the defect concentration is very large (1.4%), the transport capacity of the graphene nanopore near the Fermi level becomes very weak. Meanwhile, it can be seen that the transport capacity of the graphene nanopore structure is still reduced when the defect concentration is relatively low (0.17%). Therefore, for the graphene nanopore structure with defects, the electronic transport capacity is determined by both the defect and the nanopore structure. The influence of nanopore structure dominates at low defect concentration. With the increase of the defect concentration, the influence of vacancy defects on the transport capacity gradually gets stronger.

6.5.4 Effect of Nanopore Shape on Electrical Properties

In order to study the effect of the nanopore shape on transport properties, four different models are constructed, as shown in Fig. 6.27. In which the number of deleted carbon atoms to create the nanopore is the same in different structures. It has been reported that the distance between the nanopore and the edge could affect the quantum threshold effect [39], which causes the change of the graphene transport properties. In this study, the distance between the nanopore and the edge is kept constant to investigate the influence of the symmetry of nanopore on the transport properties. The calculation conditions are consistent with those described in Fig. 6.23.

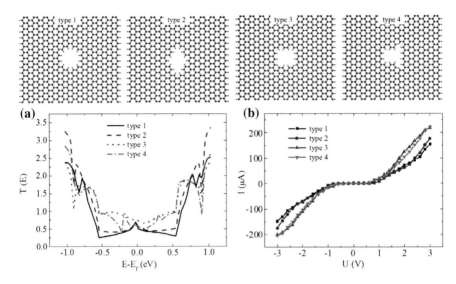

Fig. 6.27 a Transport lines at zero bias and b current-voltage relationship of graphene structure with different shapes of nanopore

Figure 6.27 shows the electronic transport properties of graphene structures with different pore shapes. Four types of structures with different symmetries were considered. Type 1 is the circular hole described in Fig. 6.23, with multiple symmetrical axis; type 2 has the elliptical pore, with a horizontal (perpendicular to the transport direction) and longitudinal (transport direction) symmetrical axis; type 3 has a symmetrical axis along the longitudinal direction; type 4 has a symmetrical axis along the transverse direction. From the results of Fig. 6.27a, it can be seen that near the Fermi level, the transport capacity of type 1 and type 2 structures with good symmetry is the weakest. With the decrease of symmetry, the inhibition of transport capacity by the nanopore structure is gradually weakened. Therefore, during the actual applications, the electrical characteristics of the structure can be adjusted by controlling the shape of the nanopore. From Fig. 6.27b, it can be seen that under the same bias, the current through type 1 structure is the lowest, and the current through the graphene structure will gradually increase with the decrease of the symmetry. This phenomenon is more obvious under high bias. For example, when the bias voltage is 2.4 V, the currents through type 1, type 2, type 3, type 4 structures are 93.3, 98.6, 168, 151 µA, respectively. Meanwhile, the current inhibition by the pore with longitudinal symmetry is weaker than by the pore with transverse symmetry.

6.6 Summary

In this chapter, the mechanism and influencing factors of nanopore processing in graphene were obtained by experiment and simulation methods. The mechanical properties and electronic transport properties of nanopore were discussed. The results show that:

1. FIB can be used to process graphene to obtain pore structure with tens of nanometer. With the change of parameters, the graphene structure will evolve from defects, nanocrystalline to amorphization. In contrast, the focused electron beam irradiation can obtain nanopore with much smaller size.
2. The fabrication of nanopore in suspended graphene is caused by direct sputtering from the collision of the irradiated particle beam and the dragging by the adjacent carbon atoms. The nanopore in supported graphene is caused by the direct sputtering, adjacent dragging and secondary collision caused by the sputtered atoms. The optimum energy and dose required for the formation of suspended and supported graphene nanopores are 280 eV, 500 and 80 eV, 200 respectively. The optimal absorption energy required for the processing of suspended graphene nanopore by electron beam was 5 eV. Meanwhile, the optimum energy increases linearly with the increase of the number of graphene layers, and different layers will be chemically bonded for multilayer case.
3. Under uniaxial tension, stress concentration occurs near the nanopore, which reduces the tensile failure behavior of graphene. The effect of chirality on the properties of graphene with nanopore is weaker than that of graphene without pore structure. The graphene nanopore is demonstrated to be mechanically immune to low concentration vacancy defects.
4. Nanopore structure can inhibit the electronic transport properties of graphene nanoribbon, and the inhibitory effect increases with the increase of nanopore size. Vacancy defects will have a further inhibitory effect on the electronic transport properties. With the decrease of nanopore symmetry, the inhibitory effect of pore structure on the transport capacity of graphene will be weakened.

References

1. Branton D, Deamer DW, Marziali A et al (2008) The potential and challenges of nanopore sequencing. Nat Biotechnol 26:1146–1153
2. Venkatesan BM, Bashir R (2011) Nanopore sensors for nucleic acid analysis. Nat Nanotechnol 6:615–624
3. Gu LQ, Shim JW (2010) Single molecule sensing by nanopores and nanopore devices. Analyst 135:441–451
4. Siwy ZS, Davenport M (2010) Nanopores: graphene opens up to DNA. Nat Nanotechnol 5:697–698
5. Wanunu M, Dadosh T, Ray V et al (2010) Rapid electronic detection of probe-specific microRNAs using thin nanopore sensors. Nat Nanotechnol 5:807–814

6. Novoselov K, Geim AK, Morozov SV et al (2004) Electric field effect in atomically thin carbon films. Science 306:666–669
7. Avdoshenko SM, Nozaki D, da Rocha CG et al (2013) Dynamics and electronic transport properties of DNA translocation through graphene nanopores. Nano Lett 13:1969–1976
8. Wells DB, Belkin M, Comer J et al (2012) Assessing graphene nanopores for sequencing DNA. Nano Lett 12:4117–4123
9. Tanugi DC, Grossman JC (2012) Water desalination across nanoporous graphene. Nano Lett 12:3602–3608
10. Sint K, Wang B, Král P (2008) Selective ion passage through functionalized graphene nanopores. J Am Chem Soc 130:16448–16449
11. Ziegler JF, Biersack JP, Littmark U (1985) The stopping and range of ions in matter. Pergamon, New York
12. Varchon F, Feng R, Hass J et al (2007) Electronic structure of epitaxial graphene layers on SiC: effect of the substrate. Phys Rev Lett 99:126805
13. Nguyen TC, Otani M, Okada S (2011) Semiconducting electronic property of graphene adsorbed on (0001) surfaces of SiO_2. Phys Rev Lett 106:106801
14. Moscatelli F, Scorzoni A, Poggi A et al (2006) Radiation hardness after very high neutron irradiation of minimum ionizing particle detectors based on 4H-SiC p-n junctions. IEEE Trans Nucl Sci 53:1557
15. Wu X, Zhao HY, Yan D, Pei JY (2015) Investigation of gallium ions impacting monolayer graphene. AIP Adv 5:067171
16. Teweldebrhan D, Balandin AA (2009) Modification of graphene properties due to electron-beam irradiation. Appl Phys Lett 94:013101
17. Abbas AN, Liu G, Liu B et al (2014) Patterning, characterization and chemical sensing applications of graphene nanoribbon arrays down to 5 nm using helium ion beam lithography. ACS Nano 8:1538–1546
18. He K, Robertson AW, Gong C et al (2015) Controlled formation of closed-edge nanopores in graphene. Nanoscale 7:11602
19. Liu S, Zhao Q, Xu J et al (2012) Fast and controllable fabrication of suspended graphene nanopore devices. Nanotechnology 13:085301
20. Li J, Stein D, McMullan C et al (2001) Ion-beam sculpting at nanometer length scales. Nature 412:166–169
21. Zhu Y, Yi T, Zheng B et al (1999) The interaction of C60 fullerene and carbon nanotube with Ar ion beam. Appl Surf Sci 137:83–90
22. Ong Z, Pop E (2010) Molecular dynamics simulation of thermal boundary conductance between carbon nanotubes and SiO_2. Phys Rev B 81.155408
23. Abramova V, Slesarev AS, Tour JM (2013) Meniscus-mask lithography for narrow graphene nanoribbons. ACS Nano 7:6894–6898
24. Jang I, Sinnott SB (2004) Molecular dynamics simulation study of carbon nanotube welding under electron beam irradiation. Nano Lett 4:109–114
25. Pregler SK, Sinnott SB (2006) Molecular dynamics simulations of electron and ion beam irradiation of multiwalled carbon nanotubes: the effects on failure by inner tube sliding. Phys Rev B 73:224106
26. Wu X, Zhao HY, Pei JY (2015) Fabrication of nanopore in graphene by electron and ion beam irradiation: influence of graphene thickness and substrate. Comput Mater Sci 102: 258–266
27. Bao W, Miao F, Chen Z et al (2009) Controlled ripple texturing of suspended graphene and ultrathin graphite membranes. Nat Nanotechnol 4:562–566
28. Merchant CA, Healy K, Wanunu M et al (2010) DNA translocation through graphene nanopores. Nano Lett 10:2915–2921
29. Lv WP, Chen MD, Wu RA (2013) The impact of the number of layers of a graphene nanopore on DNA translocation. Soft Matt 9:960–966
30. Wu X, Zhao HY, Zhong ML, Murakawa H, Tsukamoto M (2014) Molecular dynamics simulation of graphene sheets joining under ion beam irradiation. Carbon 66:31–38

31. Liu Y, Chen X (2014) Mechanical properties of nanoporous graphene membrane. J Appl Phys 115:034303
32. Pei QX, Zhang YW, Shenoy VB (2010) A molecular dynamics study of the mechanical properties of hydrogen functionalized graphene. Carbon 48:898–904
33. Zhao H, Min K, Aluru NR (2009) Size and chirality dependent elastic properties of graphene nanoribbons under uniaxial tension. Nano Lett 9:3012–3015
34. Liu L, Wei N, Zheng Y (2013) Mechanical properties of highly defective graphene: from brittle rupture to ductile fracture. Nanotechnology 24:505703
35. Ansari R, Ajori S (2012) Mechanical properties of defective single-layered graphene sheets via molecular dynamics simulation. Superlattices Microstruct 51:274
36. Gorjizadeh N, Farajian AA, Kawazoe Y (2009) The effects of defects on the conductance of graphene nanoribbons. Nanotechnology 20:015201
37. Zhang T, Li XY, Kadkhodaei S et al (2012) Flaw insensitive fracture in nanocrystalline graphene. Nano Lett 12:4605–4610
38. Kumar S, Li XY, Haque A et al (2011) Is stress concentration relevant for nanocrystalline metals? Nano Lett 11:2510–2516
39. Zheng XH, Zhang GR, Zeng Z et al (2009) Effects of antidots on the transport properties of graphene nanoribbons. Phys Rev B 80:075413

Chapter 7
Conclusion

7.1 The Main Conclusions

The application of graphene cannot be achieved without the processing of its structure. In this work, the graphene nanostructures were processed by particle beam irradiation. Based on the different mechanisms existed during the interaction between different particle beams and graphene, the methods of low energy ion implantation to dope graphene, and particle beam irradiation to join graphene and create graphene nanopore were proposed and studied. Meanwhile, the mechanisms of nanostructures processing were analyzed by atomic simulation methods. Then, the mechanical and electronic transport properties of the nanostructures were analyzed, which laid the foundation for the application of graphene. The main conclusions of this paper are as follows:

1. The damage energy threshold of graphene under single pulse laser action with different pulse widths is obtained. Under the action of the laser energy less than the threshold, the graphene structure will appear a strengthen phenomenon for the in-plane compound and out-of-plane fluctuation. When the laser energy is higher than the damage threshold, the graphene microribbon and the micro-pore can be processed. The structure of graphene under the action of ion beam irradiation will appear four kinds of phenomena with different incident energy: reflected ions, adsorbed ions onto the graphene surface, embedded ions in graphene internal and penetrated ions. The effect of secondary collision caused by the sputtered substrate atoms is always greater than that of the direct collision from incident ions. Under the irradiation of electron beam, the defects will gradually appear in the graphene structure and a nanocrystalline structure will be formed. Under ideal conditions, the failure threshold of graphene structure is 10 keV, while the vacancy defect and grain boundary structure in the actual graphene will reduce its failure threshold.
2. After low-energy ion implantation, the adsorption atoms and vacancy defects are first formed in the graphene. In the subsequent steady-state equilibrium

process, the adsorbed atoms will combine with the vacancy defects to form the structure of substitution doping or substitution doping + polygon defects. The increase in energy and dose of the ion beam leads to severe damage to the graphene structure. Under low energy nitrogen ion implantation, the optimum energy and dose of graphene doping are 60 eV and 3.125×10^{14} cm^{-2}, respectively.

3. In the case of ion doping with high concentration, the graphene tends to show a significant tensile stress concentration near the dopant atoms. Doping reduces the tensile failure strain and elastic modulus of graphene, and the higher the aggregation degree of dopant elements, the greater the decrease in tensile mechanical properties of graphene. The defects introduced during the doping will change the stress distribution in the graphene structure. The high concentration defects will greatly reduce the tensile strength and failure strain of the graphene structure. Doping with different species of elements has different effects on the mechanical behavior of graphene.

4. Implantation of doped ions will inhibit the electronic transport properties of graphene. The different forms of doping have different effects on the transport properties, and can produce transmission valley at different positions of transport line, which could induce the transition from donor and acceptor semiconductor. And the doping will destroy the symmetry of graphene structure, resulting in the phenomenon of current increase through the graphene nanoribbon structure under bias. The location of the dopant element also affects the transport properties.

5. Ion beam irradiation, laser-current combined action and thermal annealing process can form the joining results of overlapped graphene. The ion beam irradiation induced joining is due to the formation of new chemical bonds between the graphene layers, which can be induced by the "coordinated joining" of the carbon atoms in graphene and the "bridging" effect of the embedded ions. Laser and thermal annealing can only lead to the enhanced intermolecular forces between overlapped graphene layers, other than the formation of chemical bonding. Under the action of laser heat, there is no effective chemical bonding between two graphene with overlapped and T-type forms, while chemical bonding can be formed for the types of corner, butt and dislocated joint, which is due to the coordination effect of the edge carbon atoms in graphene.

6. The presence of defects in the joint makes it easy to become a tensile stress concentration site. The crystalline angle between two graphene affects the presence of defects in the joint, resulting in a localized maximum value of the mechanical properties at the crystalline angle of 0° and 60°. After butt joining of several pieces of graphene, the crack is easy to initiate at the grain boundaries during stretch. There are two kinds of fracture modes for the failure of graphene, i.e. "intergranular fracture" and "transgranular fracture". Under the irradiation of carbon ion and argon ion, the ion beam parameters to form the optimal mechanical properties of the lap joint are 1.06×10^{15} ions/cm^2, 40 eV and 1.9×10^{15} ions/cm^2, 60 eV.

7.1 The Main Conclusions

7. The electronic transport properties of graphene will change greatly after joining. Due to the difference of the orbital energy level between the different sheets and the local state of the defect at the joint, the transport capacity of the butt joint is much weaker than that of the original monolithic graphene. By controlling the crystal orientation angle between two graphene and the area ratio, the transport properties of the joint can be adjusted. The newly formed chemical bonds in the lap joint will promote the enhancement of the transport capacity, and the effect of "embedded ion joining" is stronger than that of the "coordination joining". The increase of the number of bonds between upper and lower layers of graphene would further enhance the transport capacity.
8. The use of FIB and electron beam could fabricate nanopore structure in graphene with a size of tens of nanometers and even within 10 nm. For the suspended graphene, the collision of the particle beam leads to the direct sputtering of the carbon atoms and simultaneously the "dragging" of the adjacent carbon atoms, which will result in the formation of pore structure. For the graphene supported by the substrate, besides the role of "direct collision" and "neighbor dragging", the secondary collision from the sputtered substrate atoms could also lead to the formation of nanopore structure. The optimal ion beam energy and dose required for the nanopore processing in suspended and supported graphene are 280 eV, 500 and 80 eV, 200, respectively. The optimum energy of multi-layer graphene nanopore processing increases linearly with the increase of the number of layers, and chemical bonding occurs between the graphene layers.
9. Under uniaxial tension, the stress concentration will appear in the vicinity of the nanopore, which could lead the reduction of the tensile strength of graphene, and the effect of the nanopore on the mechanical properties of graphene will be enhanced with the increase of the pore size. The effect of chirality on the properties of graphene containing nanoporous structure is much weaker than that of non-porous structure. The graphene nanopore is mechanically immune to the low concentration vacancy defects.
10. The presence of nanopore defects will suppress the electronic transport properties of the graphene nanoribbon, and the inhibitory effect increases with the increase of the pore size, which is caused by the local state on the non-frontier molecular orbital. The vacancy defects can inhibit the electronic transport performance of graphene nanopore structure, and the effect of high concentration defects is more obvious. With the decrease of symmetry, the inhibitory effect of graphene pore structure on the transport capacity is weakened.

7.2 Future Plan

The processing of graphene nanostructures strongly depends on the current development level of the equipment. In this paper, we have explored the doping, joining and nanopore processing of graphene. The experimental exploration is elementary. While the results could demonstrate the processing feasibility of the graphene nanostructure using particle beam irradiation, and lay the groundwork for future higher precision, higher quality nanostructure processing. Meanwhile, a large number of theoretical simulation works rationally explained the experimental phenomena, and detailedly described the properties of the obtained graphene nanostructure, which can also provide a reliable guide for future experimental work. Therefore, combined with the existing results of this work, the future research will focus on, on the one hand the more detailed and sophisticated experimental work, and on the other hand, the use of particle beam processing method to assemble graphene-based devices, and realize the applications of graphene nanostructures. So as to realize the controllable processing of graphene nanostructures by particle beam, and to expand the application prospect of graphene nanostructures.